CITIES
AND
SICKNESS

Volume 25, URBAN AFFAIRS ANNUAL REVIEWS

CITIES AND SICKNESS

Health Care in Urban America

Edited by

ANN LENNARSON GREER
and
SCOTT GREER

Published in cooperation with the Urban Research Center, University of Wisconsin—Milwaukee

Volume 25, URBAN AFFAIRS ANNUAL REVIEWS

SAGE PUBLICATIONS
Beverly Hills / London / New Delhi

For information address:

SAGE Publications, Inc.
275 South Beverly Drive
Beverly Hills, California 90212

SAGE Publications India Pvt. Ltd.
C-236 Defence Colony
New Delhi 110 024, India

SAGE Publications Ltd
28 Banner Street
London EC1Y 8QE, England

Printed in the United States of America

Library of Congress Cataloging in Publication Data

Main entry under title:

Cities and sickness.

    (Urban affairs annual reviews ; v. 25)
    "Published in cooperation with the Urban Research Center, University of Wisconsin-Milwaukee."
    Bibliography: p.
    1. Medical care—United States. 2. Urban health—United States. 3. Medical policy—United States.
4. Urban health—Government policy—United States.
I. Greer, Scott A. II. Greer, Ann Lennarson, 1944-
III. Series.
HT108.U7 vol. 25   307.76s   [362.1'0973]   83-15422
[RA395.A3]
ISBN 0-8039-2127-6
ISBN 0-8039-2128-4 (pbk.)

FIRST PRINTING

Contents

1. Health Care in American Cities: Dedicated Workers in an Undedicated System
   Scott Greer 7

PART I: THE STATUS OF HEALTH CARE IN AMERICAN CITIES

2. Urbanization and Health Status
   J. John Palen and Daniel M. Johnson 25

3. The Reconfiguration of Urban Hospital Care: 1937-1980
   Alan Sager 55

4. Mortality, Morbidity, and the Inverse Care Law
   John B. McKinlay, Sonja M. McKinlay, Susan Jennings, and Karen Grant 99

5. The City's Weakest Dependents: The Mentally Ill and the Elderly
   Ann Lennarson Greer, Scott Greer, and Tom Anderson 139

6. The City and Disability
   Clyde J. Behney, Anne Kesselman Burns, and H. David Banta 179

PART II: THE MAKING AND UNMAKING OF HEALTH POLICY

 Health Care in America: A Political Perspective
   Henry J. Schmandt and George D. Wendel 213

8. Politics as Accusation: New York's Public and Voluntary Hospitals
    Fred H. Goldner 245

9. Urban Health Care: Change and More Change
    Eli Ginzberg 265

10. Urbanization and Health Services: A World Perspective
    Milton I. Roemer 277

About the Contributors 299

Health Care in American Cities: Dedicated Workers in an Undedicated System

SCOTT GREER

☐ THE CITY IS A CENTER of human illness and death. It is a place of concentration, exchange, and diffusion of germs and poverty. As the place of possible cures, it creates illness; as the last hope of the hopeless, it is a citadel of death. The richer the society and the more dominant the city within it, the more these truths hold.

But the health of the city's people is one of the society's major strengths. They are privileged with the advantages of government and market, and a disproportionate share of the societal surplus is theirs. They live near the centers for the concentration, exchange, and diffusion of medical knowledge. Indeed, the industries based on health and illness are one major reason for the city's growth. From Lourdes, France, to Rochester, Minnesota, the market of the ill is an important part of the economic base of modern cities. (The best export is one for which the buyer pays all the freight.)

The cities are also killed by illness and death, from microparasites (the plague) to macroparasites (war and pillage). They are far removed from simple subsistence society and are accordingly vulnerable (McNeil, 1977). In the most profound and banal senses, the health of the city is critical to that of the society. Thus it is time for us to investigate the health policies of and for the cities, and the implications of urban policies for health and health care.

What is striking to those who have been immersed in urban studies and then have become interested in the social response to health and ill health is the extreme segregation of the two areas of inquiry. To understand all is not necessarily to forgive all, but it helps us to understand if we remember that each was a response to contemporary changes and problems in society, all ultimately produced by the increase in societal scale (Greer, 1962, 1970). The transformation of Western society by industrial development and consequent urbanization have been a generic focus of social science from Booth's *Life and Labour of the People in London* (1882-97) until the present; the process transformed much of social science and history into urban studies. At the same time, the social, physical, and cultural transformation of humanity in the process caused disruption of old patterns of maintaining health, defining illness, and caring for it. From the successes of public health in making the city habitable (through such major public enterprises as sewerage systems, garbage disposal, and police) to the mid-twentieth-century concern for the access to health care of poor central-city populations, much of the subject matter of "health studies" has been typically urban.

Yet the two strands of inquiry have been almost completely separate. Indeed, with the great achievements in public health services in the late nineteenth and early twentieth centuries, general interest in social science and the interests of those who have come to be called "urbanists" waned. Meanwhile, as the territorial state assumed greater responsibility for human welfare, including health care, the serious study of health services by social scientists increased, but relatively unrelated to the urban context.

The costs have been great to each field. The study of health and health services has tended to focus on very large units, essentially statistical aggregates ranging from nations, down to regions ("West North Central") and individual states. Then, for the United States as a whole, there are rural-urban comparisons. Large populations are differentiated by the same kinds of variables used by the Bureau of the Census, for the twin reasons that (1) these were the units available from the Census, and those used for aggregating health data, and (2) the making of policy was the focus of research, and that health policy was usually made at the national or state level. Thus the local reality within which the well and the ill subsist, the specialists and organizations who care for health needs *in this city at this time,* tends to be totally invisible.

But the metropolitan society is dominant. It is the center of wealth, knowledge, power — to prevent and cause, care for, and sometimes heal the illnesses of the human beings in the society. Seen within the metropolitan framework, health has often been relegated to one of the "free givens" assumed by economists, although health and illness have historically been critical to the fortunes of cities.

Today the provision of goods and services called "health care" consumes around 10 percent of the gross national product of the United States. In short, on the average we work well over a month out of each year for health care. On the basis of this market and its provision, a very large part of the city's work force is dependent, from physicians who are wealthy and have access to governors to nurses' aids who carry bedpans in the hospitals. Moreover, the organizations in this arena may compete bitterly, as in the running battle among professions or the fight for survival and dominance among hospitals. To see health, illness, prevention, and care outside the context of urban society is a sterile view.

But the urban sociologist, political scientist, or economist has had a complementary blindness. He rarely gives due weight to the social reality of the health care segment of urban organization. When he does, he tends to treat it as exogamus, given. In the process, he is apt to accept the conventional wisdom, even though it is a compound of biased observations, sentimental and outdated clichés, propaganda put forth by obvious interest groups. As with religion, for which social scientists ordinarily have a "tin ear" according to Max Weber, the observers tend to be more "social" than scientist. Perhaps they are victims of their culture's general avoidance of such topics as old age and illness, death and insanity.

To be sure, there has been a substantial investment of social science resources in the health segment of the society. At one point the sociology of health and medicine was the fastest-growing specialty in the discipline (Goode, 1973: 22). However, this work had very little effect on the general thinking of sociologists. The reason may go back to its emphasis, derived from its public funding. In Eliot Freidson's phrase, it was typically "sociology in medicine" rather than "sociology of medicine." Freidson's study, *Professional Dominance* (1970), emphasized not only physicians' political control of the health work force, but also their definition of the situation. The medical model, above all, emphasizes the centrality of the physician as solo performer in one-on-one health care. Consequently, sociology and the other social sciences, insofar as they accepted this definition, were one more cadre of physicians' aides.

Mainstream social scientists continued to think of the general area of health as a given, medicine as a consumer good. Others, however, were investigating the history of this enormous cultural complex and organizational segment of modern society, with enlightening results. Rosemary Stevens (1971) documents in detail the transformation of a large number of ill-trained, poorly paid claimants to the title of physician into a cadre of "specialists" who tightly controlled the entry point into the work force, dominated hospitals, and achieved a monopoly on new technical instruments (see also Bucher and Strauss, 1961). Other social scientists became intrigued with the demographic paradox. Although modern English society rests upon a great increase in population dating back several centuries, and massive improvements in survival rates and health have had much to do with the flourishing of England, medicine as we know it had nothing to do with the process (McKeown, 1976). Obviously, the kind of medicine identified by Freidson as "dominant" had little claim on many of the modern improvements in human health often allocated to its credit.

Such inquiry might have been developed by urban demographers and ecologists, but was not. However, one sociologist, Fred Cottrell, published a profound discussion of the human being as a species-in-environment almost thirty years ago. *Energy and Society* (1955), the right book at the right time, was not read by the right people. Cottrell was concerned with the biological base of human life, its dependence on technological mediation and nonhuman energy sources for the huge population increase we have seen, and its ultimate vulnerability, for the same reasons that concerned Malthus. Most important, for the subject of this volume, Cottrell underlined *the double nature of the human being as a social animal.* On the one hand, we live by transactions with the environment, the sweat of our brow and the cunning of our tools: We may be seen as the "human resources" which spokespersons, leftist and rightist, see as a factor of production. On our labor rests the success and even the persistence of the enterprise. But Cottrell points out our other nature — as consumers, fellow beings who are entangled in complex rights and duties, citizens — a value in itself. As such, we are given claims on the surplus of the society by virtue of being born and raised to be human.

This is the source of continuing dilemmas. Specifically, how much of the society's surplus shall be consumed by nonproducers, through economic support, medical support, and the like? How much shall be distributed to them during times of scarce resources? One way or the other, societies must answer the questions in such a way as to persevere, or go under. In poor societies, infant mortality has been a delayed and costly form of birth control. In our society, with a wealth of material goods and technologies, we still face the basic questions: Who shall survive, who shall reproduce, and how much under what circumstances? The way contemporary large-scale societies will make such grave policy decisions we do not know.

As the worldwide euphoria that followed World War II dissipates, for the best of economic and political reasons, the limits of our technology become apparent. With them comes a sobering reassessment of technologies and policies. In the study of health services this is manifest in the work that has followed Thomas McKeown's analysis of England's population growth as related to medicine, including work by McKinlay et al. (some of which follows), and by a worldwide interest in birth control as well as production increases. With respect to the social study of medicine and health care generally, a more critical (and creative) stance has developed from the pioneering work of Anselm Strauss and associates, Howard Becker and

Eliot Freidson (Becker et al., 1961; Bucher and Strauss, 1961; Schatzman and Strauss, 1966; Freidson, 1970).

Sociologists studying health services are now more apt to ask "What's going on here?" than "How can I help you?" The occupational groups and the social organization of health services are objects of inquiry in the same frame of references as other occupations, organizations (Perrow, 1978; Greer, 1983; Wildavsky, 1979: Ch. 12).

This approach is also manifest in the work of the medical criticism of modern medicine. Much of it evolved from epidemiology, an application of social science techniques to the study of medical effects and health in general; indeed, McKeown's work stems from this source. The use of random clinical trials, for example, should edit many unnecessary practices from standard medicine (Cochran, 1972). As the nature of life, and the human as a biosocial being in an environment, is more generally aprehended, the notion of one specific etiology for each illness declines (DuBos, 1959: Ch. 4).

Meanwhile, the hard truth has emerged — that, much of the time, all we have available is "halfway medicine," care that helps maintain the organism but cannot cure the illness (Thomas, 1974: "The Technology of Medicine").

In the following chapters, scholars from a variety of disciplines address the implications of urbanization for human health and the prevention and care of illness. The authors, from the disciplines of economics, sociology, political science, and health planning, were asked to answer complementary questions: If you are an urban sociologist (for example), what does the health and health care of the city's population signify to you? Or (if you are a health specialist) what does urbanization, and the city as a matrix for life and interaction, mean to the study of health and health services?

Our authors concentrated, with one exception, on urbanization and health in the United States, with a heavy emphasis on the contemporary situation and the history from which it evolved. The sections of the book are, as the reader will find, somewhat arbitrary in that certain themes and problems turn up again and again. But we have divided the volume into two sections: The Status of Health Care in American Cities and The Making and Unmaking of Health Policy.

THE STATUS OF HEALTH CARE IN AMERICAN CITIES

In "Urbanization and Health Status," J. John Palen and Daniel M. Johnson explore the historical relationship between cities and health status, and the way it has changed as the society increases in scale and affluence. They then turn to contemporary rural-urban comparisons, noting the historical changes and general equalization of rates taken to measure health as we approach the present. They note the weakness of our measures for individual cities (and, we might add, the scarcity of city studies), specifically the absence of any generalized indicator of health status. All point toward that rift between urban studies, on the one hand, health studies, on the other.

Appraising the changes in the distribution and structure of American populations together with advances in communications, they hazard the guess that the historical concentration of medical resources and the best medical skills in cities may be coming to an end. Modern technology, in their view, makes possible a much more general dispersion of the population, and instant communications make possible, in principle, the "wired society" in which the expert consultant and sophisticated analysis are available far from the home office.

Alan Sager is troubled by the unplanned life and death of hospitals in the urban complex. In "The Reconfiguration of Urban Hospitals: 1937-1980" he investigates closings and survival in fifty-two large and middle-sized cities, the hospitals studied accounting for almost one-fourth of all acute care hospital beds in the United States. From his analysis it is clear that the hospitals that closed were no more "redundant" than those that remained in business, but they *were* more apt to care for the poor and the ethnic populations of the inner city, and they were disproportionately dependent on such funding as these patients could bring. As Sager points out, the demise of such hospitals, with their lack of research and teaching functions and their emphasis on care, leaves for the inner-city population only the "medical centers" that have survived through such activities. These, in turn, provide the most expensive hospital care in the society, while the hospitals going under were among the cheapest.

As Sager demonstrates, the private hospitals that invest heavily in high technology are not the villains of the piece: They must do so in order to survive, given the way we fund our hospitals. Thus his solution is one that transcends specific advantages, threats, bribes, and regulations. The entire playing field must be changed and some policy devised to create an overall order that provides "universal financial access" for the citizens and a reasonable security from closure to hospitals. He discusses some existing state policies that promise to produce these results. The logic of his argument seems conclusive, but, since those who will suffer most from a continued working out of the dynamics of "competition" are the poor of the inner city and the hospitals who serve them, it is by no means certain to prevail politically.

John B. McKinlay, Sonja M. McKinlay, Susan Jennings, and Karen Grant address a major assumption of most health care studies — that modern medicine is increasing the average life span and that there is a continuous improvement in the health status of the population blessed with such medicine. "Mortality, Morbidity, and the Inverse Care Law" is an extension of Thomas McKeown's work in England to the contemporary American population, and demonstrates the negligible effects of modern medicine on the increase in life expectancy we have experienced in the twentieth century. Furthermore, using a measure of "impaired activity", McKinlay and his colleagues suggest that what trivial extension of life has been achieved on the average has, on the average, been an extension of illness. At the same time that we are investing enormous resources in the "standard medical practices" that do little to maintain health and life, we are skimping on care for those who need it most. The inverse care law says just that, reinforcing Sager's conclusion that we neglect the most needy, who would also repay most easily the investment of care, while spending our resources far beyond the point of diminishing returns on the more medically prosperous.

This eloquent argument has a number of policy implications beyond this, of which the most important, perhaps, is the admonition that it is unjust to spend public funds on medical practices of little or no proven value — particularly in view of the lack of the simplest proven help for much of the population. While McKinlay and his colleagues do not regard modern medicine as a total waste, it is clear from their analysis that much of it is not *medically* cost-effective.

Palen and Johnson make it clear that, for whatever reasons, the longevity of the American population is increasing; McKinlay and his associates emphasize the decline in death from acute and contagious ailments, and the commensurate increase in death from degenerative diseases, brought on by the biological "wearing down" of the organism. Indeed, many of the medical waifs of the inner city whose care concerns Sager are older people with degenerative ailments. In "The City's Weakest Dependents: The Mentally Ill and the Elderly," Ann Lennarson Greer, Scott Greer, and Tom Anderson describe and analyze the care of such persons in American society. Beginning with a historical sketch of their fate as modern society emerges, the concern and lack of concern evinced in earlier and much poorer societies, they detail the American history. This begins in dispersed and unaccountable care by kin and the local community (if any), succeeded by the gathering of the mentally ill, in particular older "senile" persons, into state institutions for more responsible care. The movement for community care, sponsored by advocates who were optimistic about therapy for the mentally ill and aged, by civil rights advocates and courts, and by fiscal conservatives who wanted to save state taxes, came to its fruition in the 1960s. The results, community mental health centers for a large part of the population together with the drastic reduction of the institutionalized population, are described and analyzed. What emerges is the presence of a residual population of the incurable helpless, who, under the new regime of "freedom from restraint" and "community care," often get no care at all. We appear to be, in some respects, more uncaring and irresponsible than ever. A key question is now this: Will the asylum, the rural institution sequestering the decrepit old, the mad, and other social misfits, be reconstituted? These populations are a costly responsibility, yet their care is a harbinger of a population living longer and, inevitably, joining their ranks sooner or later.

In a continuation of this general discussion, Clyde J. Behney, Anne Kesselman Burns, and H. David Banta consider the city as an environment for those less than normally able to function. "The City and Disability" begins with an analysis of current thought on the subject, clarifying the distinction between a physical or mental *impairment,* its translation into a functional *disability,* and the consequent *handicap* the individual carries in social life. They note that the handicap may be lightened in various ways, by addressing the disability (as the community mental health reformers hoped to do), by technological, ecological, and other devices that diminish the effects of the disability (from eyeglasses to artificial limbs or voice), and by making the handicap less relevant through modifying the environment (conditions of travel and work) and social definitions (rigid and often quite arbitrary requirements) of roles.

Their essay underlines the great freedom that the human being has from the environment, through the technologies with which we modify the meaning of that environment. If we can exist in outer space and under the Antarctic ice cap, surely we can perform useful roles, live acceptable lives, with a large number of the impairments that in the past relegated one to the status of beggar, cripple, or second-rate citizen.

THE MAKING AND UNMAKING OF HEALTH POLICY

Henry J. Schmandt and George D. Wendel underscore the lack of interest among political scientists in the enormous complex of activity called "the health system." "Health Care in America: A Political Perspective" is an effort to chart the structure of controls in this vast segment of our social life. It is marked, as much as anything, by the antagonistic cooperation of groups with both differing interests and differing principles of order. The citizen who is ill, or in hazard of illness, is the responsibility of the public health service, at federal, state, county, and municipal levels. He or she may go for relief to private-office physicians, clinics of a public, private, or public not-for-profit sponsorship, hospitals called private voluntary, or county, city, or Veterans Administration facilities. With the exception of the last, health provision is typically paid for by private and public funds, or by insurance firms, such as Blue Cross, that are defined as private not-for-profit. The care of this person will be monitored by peculiarly hybrid organizations in which the public and private interests are comingled: Boards of physicians will monitor physicians' work, in the public interest.

Thus the health service industry is one in which the dichotomies of our society, between public and private, between national and local, between professional and lay authorities, come together in very complex fashion. The long history of this development, presented in part by Rosemary Stevens's (1971) history, has not been one of clear-cut differentiation of function and ordering of rights and duties. Instead, the delivery of health services has been an arena for conflict between major antinomies of the society.

Cities as governing entities are health providers, funders, refuges of last resort. They are also economic ecosystems that are heavily invested in the health services industry, and the inner cities are increasingly dependent on that industry for jobs as well as help for illness. Yet the same cities are economically sick, and their most prosperous populations are going or gone to the suburbs; the "fee for practice," "private" physicians and hospitals are following, as Alan Sager has described.

Schmandt and Wendel consider possible futures briefly, suggesting that the political culture and the orientation of influence in the society prohibit both a universal health care system and an unbridled free market. The citizens are suspicious of the overarching system, with universal access and public control. But neither private practitioners and hospitals nor advocates of public aid for the needy favor a free market; the first have come to rely on the existing public controls for their private welfare; the last do not trust unregulated charity to repair the inequalities guaranteed by unregulated free trade. Thus the complex and messy area of health politics is apparently with us for the future.

Schmandt and Wendel develop a relatively "cool" analytical description of health politics. In "Politics as Accusation," Fred H. Goldner discusses, at some length and in some heat, what it is like to try to do the public's business in a society where politics is indiscriminately used as a damning epithet, applied differentially to those working for the public, while the equally self-serving combinations and manipulations of largely public-funded "private" parties are considered pure of such a stigma. In Goldner's New York City experience, this handicap is augmented by the glare of the mass media, who view public affairs as providing the daily entertainment that attracts viewers and readers to advertisements. With such sources of information, the average citizen is exposed to a continuing hurly-burly in which all actors and all actions are finally "political."

Under the circumstances, it is remarkable that any coherence at all results in health planning and the distribution of health resources. Indeed, what coherence there is seems largely brought about by the practitioners, from specialist physicians to ward attendants and nurses' aides, who have at least the discipline of the job to order their behavior. For the larger picture, the investments on behalf of the society as a whole, there is no glimmer of order. Reasonable and effective efforts to provide adequate hospitals with a sensible division of labor are no more evident at this sidewalk (or corridor) level than in the statistical analysis of Alan Sager.

Looking back over the past twenty-five years of health care history, Eli Ginzberg highlights the enormous changes that have been effected by federal governmental policy. Regardless of the relative impotence of our recent efforts to *cure* illness, Ginzberg's point is the great extension of *care* for the ill, particularly those least privileged by the distribution of power and wealth in society. He notes that social insurance schemes, Medicare and Medicaid, have reached many who in the recent past had only such charity as was available from public hospitals, for-profit physicians, and privately controlled hospitals. In

Ginzberg's judgment, most of the elderly who should be in nursing homes with full-time surveillance and care are there, and while he does not discuss the mentally ill, it is likely that on the whole there is more awareness of their plight, in more parts of the country, than existed before our effort to create a Great Society was exhausted.

However, in "Urban Health Care: Change and More Change," he points out that this very success has also resulted in losses. The "charity complex" has atrophied, due to the assumption of responsibility by the public in general. Moreover, with the spiraling costs of health care, due in part to publicly funded medicine, the philanthropic dollars remaining do not go far in meeting the bills hospitals must pay today. Now, with the economy in a depression and a panicked Congress and administration "biting the bullet" of relative austerity, those who will lose the most are undoubtedly those with least political strength. They are the patients dependent on the states and local governments, through Medicaid, General Assistance, or other programs not falling under the rubric of social security. These turn out to be the population discussed in "The City's Weakest Dependents," as well as the large inner-city populations without "safety nets" and, of course, those who are unemployed and have lost job-dependent insurance coverage. Thus the provision of public support for health care has created a dependence on public funding at the federal level; with a "hard-boiled" austerity movement coming down from Washington, the poor pay the most. The inverse law holds with a vengence, as McKinlay and his associates point out.

"Urbanization and Health Services: A World Perspective" is a return to the global perspective that Palen and Johnson use to introduce the volume. Milton I. Roemer, on the basis of his broad experience, emphasizes both the variety of health services in different nations and the common effects of urbanization. While the differences in health service quantity and quality may be enormous between nations, the difference between city and hinterland is everywhere dramatic, with the city the center of the most advanced health services and the concentration point for the relevant institutions. Roemer notes the sharp differences made, within the "industrial-agricultural" nations dichotomy, of a society's organization on an "entrepreneurial" versus a "socialist" basis, but everywhere there remains the dichotomy of city and country.

While there is more chance for a broader sharing of health care resources, a more evenhanded distribution of care, in the socialist societies, Roemer points out that there is also a greater danger of insensitivity to local needs and carelessness with respect to local response in the more centralized system. But carelessness about the consequences of the "system in operation" seems to be an earmark of health services, wherever they are delivered. In the title of a recent article, Aaron Wildavsky has summarized the way the United States responded to health services even before the recent economic slump. In his phrase, we are "doing better and feeling worse" (Wildavsky, 1979). In part, this may reflect the sheer cussedness of human beings when they try to operate in a system typified by fragmented interests, viewpoints, cultures, and a resulting antagonistic cooperation. Perhaps it is partly because we expect more from health care than it can provide.

WHY DO WE PAY THE BILL?

Health and health care seem to generate intense interest across a wide spectrum of the population. Most daily newspapers have full-time staffs devoted to health and medical news; weekly journals and the electronic media are certain to program discussions of health, illness, and medicine with regularity, whether or not there is any "news." A correspondingly wide segment of the society is involved as part-time and volunteer helpers, advocates and activists.

Health and illness, medicine as science, and medicine as therapy seem much more newsworthy than such traditional concerns as religion. Indeed, one may speculate that, in an increasingly worldly society, health care serves some of the functions of a secular religion. The fact that it is supposed to be based on science, and therefore is supposed to exist in another domain, is moot. The scientific basis of health and health care is as unknown to the average health consumer as were the equally abstract and precise theories of Thomistic Christianity. Nor are the proofs more accessible and convincing than those of revealed religion. In short, modern medicine rests on a profound faith in esoteric knowledge and the laying on of hands by those versed in the doctrine.

Thus the doctrinal disputes among practitioners and theorists are to be expected, as are the continual accusations that the patients do not obey the prescriptions of their wiser consultants. To be sure, some people who are devoutly religious are also believers in scientific medicine; in anthropological literature this is known as "syncretism," the process by which alien doctrines are added to the established core of belief without regard to logical coherene (voodoo, a mixture of tribal animisms and Christian animisms, is a well-known example).

Under these circumstances, health services may be seen as the efficacious branch of an organized secular religion devoted to life in this world. From this perspective it is not surprising that our American society spends 10 percent of its gross national product on this institution and its practitioners. The devout Mormon pledges a tithe of 10 percent of his income, as did the medieval Christian, plus one day's labor for God for each ten days of work for himself. While the holdings of the Mormon Church are not known to me, it is generally agreed that the Christian church, at the height of its worldly stewardship, held approximately one-third of the arable land of Europe.

However, while one can certainly see the continued thread of religion in our definitions of ill health — with illness a lapse from grace, mental illness a possession by evil spirits, and death a form of damnation — something is missing. We lack the paramedical backup of belief and faith in the will of God. All good religions rely heavily on fatalism and, for many of the outcomes of medical practice we discuss in this volume, such fatalism is a necessary completion. Talcott Parsons (1951) used to discuss religion's importance to social structure as a device for "pattern maintenance," for the social stability that allows the division of labor to work. It helps people to accept their roles, their rights and duties, and conform to them with vigor. In other words, religion "made sense out of human existence."

This is obviously beyond the scope or ambition of medicine but, in the vacuum left by the death of the gods, scientific medicine is certainly one of the belief systems developed to quell anxiety and reassure the client that life is meaningful and there is help for pain.

Thus treatment, cure, prognostication — all function as reassurance. The demand for health services is powerful and well-nigh universal; no debunking of specific practices will persuade the sick to forgo it. Expensive as they are, placebos, halfway therapies, simply the laying on of hands are, along with truly effective "scientific" medicine, in heavy demand by the citizens of the modern city.

REFERENCES

Becker, H.S., B. Geer, E. C. Hughes, and A. L. Strauss (1961) *Boys in White: Student Culture in Medical School.* Chicago: University of Chicago Press.

Booth, C. (1882-1897) *Life and Labour of the People in London.* London: Macmillan.

Bucher, R. and A. Strauss (1961) "Professionals in process." *American Journal of Sociology* 61, 4: 325-334.

Cochran, A.L. (1972) *Effectiveness and Efficiency.* London: Nuffield Provincial Hospital Trust.

Cottrell, F. (1955) *Energy and Society.* New York: McGraw-Hill.

DuBos, R. (1959) *Mirage of Health.* New York: Harper & Row.

Freidson, E. (1970) *Professional Dominance: The Social Structures of Medical Care.* Chicago: Aldine.

Goode, W.J. (1973) *Explorations in Sociological Theory.* New York: Macmillan.

Greer, A. (1983) "Medical conservatism and technological acquisitions: the paradox of hospital technology adoption," in J. A. Roth and S. R. Ruzek (eds.) *The Social Impact of Medical Technology.* Greenwich, CT: JAI.

Greer, S. (1962) *The Emerging City.* New York: Macmillan.

Greer, S. (1970) *The Urbane View.* New York: Oxford University Press.

McKeown, T. (1976) *The Role of Medicine: Dream, Mirage, or Nemesis?* London: Nuffield Provincial Hospital Trust.

McNeil, W. H. (1977) *Plagues and Peoples.* Garden City, NY: Doubleday.

Parsons, T. (1951) *The Social System.* New York: Macmillan.

Perrow, C. (1978) "Demystifying organizations," Ch. 5 in R. Sarri and Y. Hasenfeld (eds.) *The Management of Human Services.* New York: Columbia University Press.

Schatzman, L. and A. Strauss (1966) "A sociology of psychiatry: a perspective and some organizing foci." *Social Problems* 14 (Summer).

Stevens, R. (1971) *American Medicine and the Public Interest.* New Haven, CT: Yale University Press.

Thomas, L. (1974) *The Lives of a Cell.* New York: Viking.

Wildavsky, A. (1979) *Speaking Truth to Power.* Boston: Little, Brown.

Part I

The Status of Health Care
in American Cities

2

Urbanization and Health Status

J. JOHN PALEN
DANIEL M. JOHNSON

□ ONE OF THE IRONIES of our era of extensive medical research and elaborate health care delivery systems is that there is yet no universal consensus on the precise definition of what constitutes health. However, one pattern that has been observed since data on health status first became available is the extent of differences between the health status of those living in cities and those living in the country. Over the last century the nature of urban-rural differences has undergone dramatic transformation — a transformation that this chapter will explore. In so doing we hope to highlight some of the policy issues involved.

This chapter can be roughly divided into three sections. The first examines the historical relationship between cities and health status. We note, for example, that nineteenth-century cities were, indeed, "the graveyards of countrymen," having higher levels of disease and epidemics than rural places. How and why this changed is discussed.

The second section examines contemporary rural-urban patterns. Various attempts to define and measure health status, including "objective" measures as well as "subjective" self-assessment rates, are discussed. Current urban-rural rates for the various measures are presented and their limitations noted.

Finally we look toward the future and speculate on how emerging demographic patterns of dispersion, and break-

throughs in communication technology, might affect the current
pattern of urban specialists and rural practitioners.

UNHEALTHFUL CITIES

One of the basic assumptions in medical epidemiological
research is the sociological dictum that a relationship exists
between health status and spatial location. Until this century,
the most obvious example of this postulate was the differential
in health status between city residents and their rural counterparts. Empirical research dating back to John Graunt's
seventeenth-century study, *Natural and Political Observations Made upon the Bills of Mortality* (1662), established that
cities grew only through the continued immigration of rustics.
As late as 1790, the city of London was still recording three
deaths for every two births (George, 1964). The engravings of
Hogarth and the novels of Dickens viscerally convey what it
meant to live in an era in which city dwellers, particularly the
poor, were forced to subsist under brutal and unhealthful conditions.

However, disease-ridden cities were not a prerogative of
older cultures and continents only. The New World, in this as in
other areas, copied the Old. Typhoid, thyphus, tuberculosis,
and dysentery were periodically epidemic in the Americas. The
Spanish colonies largely "solved" their native problem by ex-
terminating the Indians through European-introduced
epidemics, and North American Indians fared little better.

Nor were groups of European origin immune. Philadelphia,
the largest and in many ways the most advanced city of the new
United States, lost 4,000 persons or almost 10 percent of its
population in the yellow fever epidemic of 1793. Cholera
epidemics ravaged America's cities in 1832, 1849, and 1856.
Even in "healthful" America these purges were viewed as a
more or less inevitable consequence of city life:

> Smallpox, scarlet fever, measles, diphtheria were domestic
> pestilences with which the people were so familiar that they
> regarded them as necessary features of childhood. Malarial

fevers . . . were regularly announced in the autumnal months as having appeared with their "usual severity." The "white plague," or consumption, was the common inheritance of the poor and rich alike. With the immigrant came typhus and typhoid fevers, which relentlessly swept through the tenement houses, decimating the poverty-stricken tenants. At intervals, the great oriental plague, Asiatic Cholera, swooped down upon the city with fatal energy and gathered its enormous harvest of dead. Even "yellow fever," the great pestilence of the tropics, made occasional incursions [Smith, 1911].

Some even professed to see certain advantages in the periodic reduction of population. As Thomas Jefferson wrote to Dr. Benjamin Rush, "The yellow fever will discourage the growth of great cities in our nation, and I view great cities as pestilential to the morals, the health, and the liberties of man" (Lipscomb and Bergh, 1904: 173). To balance the equation, it should also be noted that at the same time Jefferson was penning this statement, he was also advocating the planning of cities in a more dispersed pattern to reduce contagion.

Others saw disease as God's just punishment. Since the victims disproportionately came from poor central-city tenement dwellers, it was possible for more affluent citizens to consider diseases such as cholera the "Scourge of God which would strike only those who were filthy, intemperate, and immoral" (Glaab, 1963: 116). It was convenient to argue that one should not interfere with God's will. Certainly, slums had two or three times the mortality of other urban districts.

The wealthy, however, were not as immune to either the physical or the economic consequences of epidemic or disease as they might hope. This was because the physical and social ecology of the nineteenth century was one in which rich and poor shared a limited physical space. Nineteenth-century American cities remained exceptionally compact, rarely extending more than one hour's walk from the center. The industrialization of the late nineteenth century further emphasized concentration and centralization. A limited transportation technology meant workers of necessity lived in high-density — and usually filthy — tenements near the centrally located factories.

Even middle-class districts, while far more pleasant, also had to be centrally located and relatively densely inhabited. The average New Yorker at the turn of the century still lived within two blocks of his place of employment, and even in Chicago, then the prototype of the sprawling city, one-half of the inhabitants lived within 3.2 miles of the city center. Such compact cities provided excellent environments for unfettered contagion. Ironically, the great Chicago Fire of 1871, which is commonly treated as one of the world's greatest urban disasters, actually in the long run probably saved lives by leveling 1,700 densely packed acres, many of which were built up with fetid shanties and slums.

PUBLIC HEALTH AND REFORM MOVEMENTS

The unhealthful condition of American cities led, particularly during the second half of the nineteenth century, to nascent public health services, the building of municipal water systems, and more systematic sewage and garbage removal. New York, for example, took control of its sewage disposal in 1849 and within twenty years had laid almost 200 miles of pipe; Chicago constructed its first sewer in 1856, and, in fifteen years, had built 140 miles of sewers (Brown, 1967). Such measures produced a clear downward trend in infectious diseases (see Figure 2.1). Massachusetts, one of the few areas in which long-term data exist, is illustrative. In 1850, male life expectancy was 38.3 years; this had increased to 46.1 years by the turn of the century (U.S. Bureau of the Census, 1975). Life expectancy for the total U.S. population was 47 years.

The benefits of public health and related sanitation improvements, while designed largely for the upper half of the population, also benefited the poor and immigrants, who had the highest morbidity and mortality rates. Such improvements, though, were hardly the consequence of upper-class altruism. Rather, reducing the scourges of the poor also conferred sig-

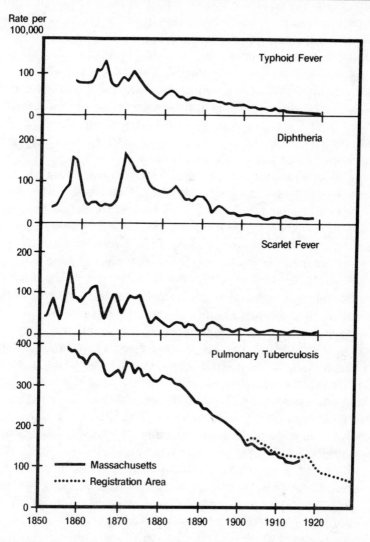

SOURCE: Sydenstricker, E. *Health and Environment.* Copyright © 1933 by McGraw-Hill
Book Company. Reprinted by permission.

Figure 2.1   Trends in Mortality from Typhoid Fever, Diphtheria, Scarlet Fever, and
Pulmonary Tuberculosis from about 1850-57 to about 1920, Massachusetts

nificant benefits on the nonpoor. Many of the affluent had come to realize that an epidemic starting in the poorer wards could not be easily contained. Increasingly, as knowledge of germ theory became common, the well-to-do recognized that the servants who cooked their meals, washed their clothing, and cared for their children could also transmit typhoid, tuberculosis, or dysentery. Thus, it was in the interest of the wealthy to provide all areas of the city with safe drinking water.[1] It was also not overlooked that a healthy populace also meant a factory work force that remained on the job.

During the latter part of the nineteenth century and the first decade of this century, public health advocates were united in a loose coalition that included liberal Protestant clergy preaching the social gospel and civic reformers. This Progressive Movement commonly linked public health improvements to overall social reform and betterment. Reformers such as Jane Addams and Jacob Riis did much to connect, in the public's mind, improvement in sanitation and health care with a defense of the home and democratic society. Elimination of disease and the slums in which it bred were seen as necessary not only for physical health but also to prevent anarchy, rioting, and the corruption of the democratic processes. As Riis (1982) put it, "You cannot let men live like pigs when you need their votes as freemen; it is not safe." Public health programs were expected to make inner-city residents not only healthier but also "better." The implicit assumption was that a physically healthy city population would also be a morally healthy population.

That the poor might not want to be reformed, and might even fight vaccination programs and the like, was simply taken as further evidence of the need for external reformers. Tenement laws, pure food and drug legislation, and settlement houses were all expected not only to improve the city populace's physical well-being but also to uplift the poor's moral character. Elimination of disease and slums — they were often paired — was expected to curb alcoholism and wife beating, encourage schooling, and promote good citizenship.

REFORMING THE IMMIGRANTS

Pre-World War II discussions of disease and unhealthfulness in the city were most often discussions of the conditions prevailing in the inner-city slums housing immigrants (Strong, 1891). By the turn of the century, over half the population of most of the larger cities of the Northeast and Midwest was of foreign stock, while the port cities of New York and Boston were over three-quarters foreign stock.[2] Massive immigration into the industrial centers — approximately one million a year during the first decade of this century — resulted in newcomers often being forced to live under degrading conditions in disease-ridden slums. At the turn of the century, the lower East Side of New York had the highest density ever reported for a Western nation — twelve blocks had between 1,000 and 1,400 persons per acre. By comparison, the most densely populated borough in London had a far fewer 182 persons per acre.

In classic blaming-the-victim fashion, the immigrants were held responsible for the conditions within which they were forced to live. High morbidity and mortality were ascribed not as much to deplorable working and living conditions as to the immigrants being of inferior biological stock (Grant, 1924). E. A. Ross, one of the leading sociologists of the era, expressed this view:

> Steerage passengers from a Naples boat show a distressing frequency of low foreheads, open mouths, weak chins, poor features, skew faces, small or knobby crania, and backless heads. Such people lack the power to take rational care of themselves; hence, their death rate in New York is thrice the general death rate [Ross, 1914: 113].

At the same time, tenement dwellers were criticized for being weak and inferior and were contradictorily portrayed as having special, almost mystical strengths against disease. As expressed by a physician, "The Slavs are immune to certain kinds of dirt. They can stand what would kill a white man"

(Ross, 1914: 291). Given that such attitudes were held by substantial portions of the scientific and medical communities, it is
surprising that mortality rates in inner-city immigrant
neighborhoods were not higher.

As might be expected, health problems were even greater
among urban blacks. At the turn of the century, the life expectancy of the black population was only 33 years, some 15 years
less than that for the white population (U.S. Bureau of the
Census, 1975: 55). Ironically, while southern slaves of earlier
generations had received at least some care, since they represented economic investements, northern industrialists had
only the most marginal concern for high mortality rates among
the black urban proletariat. Not until the social legislation of
the New Deal was national attention given to improving the
health of the black poor.

TOWARD METROPOLITAN AND
NONMETROPOLITAN CONVERGENCE

In spite of improvements made in sanitation and sewage
disposal (New York, at the turn of the century, had some 5,000
street sweepers; a health, as well as an aesthetic, necessity,
since each horse in the city was estimated to produce 22 pounds
of manure daily), urban mortality rates in the United States
remained stubbornly above those for rural areas. It is generally
agreed that this was a consequence of heavier mortality among
city children. The infant death rate for the urban population of
the registration area in 1890 was 243.3 per 1,000, as opposed to a
far lower 121.2 per 1,000 births in rural areas (Lampard, 1973:
25)

Only after the turn of the century did the cumulative effect
of pure water, better diet, more stringent inspection of milk,
isolation of contagious diseases, and inspection of schoolchildren begin to have a major impact. Infant mortality appeared
even more intractable, but by 1910 there was a clear pattern of
declining infant death rates. Figure 2.2, showing the sharp

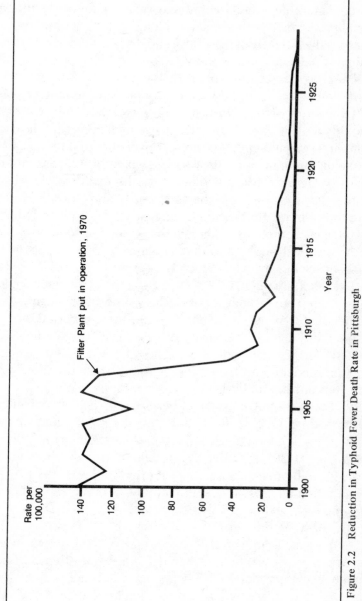

Figure 2.2   Reduction in Typhoid Fever Death Rate in Pittsburgh

SOURCE: Sydenstricker, E. *Health and Environment.* Copyright © 1933 by McGraw-Hill Book Company. Reprinted by permission.

reduction in deaths due to typhoid fever in Pittsburgh following the completion of a 1907 water filtration plant, clearly illustrates the considerable effect technology had in reducing urban mortality. Public health measures, along with technology, were changing the cities from unhealthy and disease-ridden places to comparatively safe places to live.

Thus, as the century opened, the city began to shed its reputation as an inherently unhealthy place. The pattern was changing from one of cities having universally higher urban mortality and morbidity rates to one in which the size of the city population was a factor. The greater exposure to infection and contagion found in large cities was now being compensated for in part by better medical and public health facilities. Between the two world wars, the highest morbidity and mortality rates were found in medium-size and smaller cities — places which had large city problems of contagion without the compensation of the medical-public health advantages of the large metropolises (Sydenstricker, 1933: 65).

The federally funded WPA projects of the 1930s evidenced a massive effort to provide many of these cities with clean water and adequate sewage systems. Also, during this period, urban hospital and medical delivery systems were considerably expanded. However, as the cities brought water and sewage problems under control, outlying locations began to experience more urban-like public health and sanitation problems. Partially, it was a question of numbers. As of 1920, the United States had only some 18 million persons (17 percent of the population) living on the periphery of the nation's cities. Sixty years later (1980), suburban metropolitan area dwellers numbered 103 million. While wealthy suburbs fared well, the growth of sanitation and public health facilities in other, less affluent outlying areas did not always keep pace.

The post-World War II period was noted for mass suburban development. Many of the new housing subdivisions were constructed as essentially urban densities, but often well beyond urban amenities such as existing water and sewage systems. In the early years, a subdivision's reliance on private wells and septic tanks may have been marginally adequate.

However, as housing filled in and vacant lots disappeared, problems not infrequently arose with soil percolation and water contamination.

The postwar period also witnessed problems not directly associated with housing. In rural areas there has developed a growing problem of agricultural chemicals, particularly nitrates, leeching into water systems and affecting the food chain. Rural areas also have solid waste problems. Cattle feed lots in Iowa, for example, produce more animal wastes than do the people of any city of that state, but without the elaborate sewage treatment systems of the cities. Rural areas also provide the landfills in which cities and industries dump their solid, chemical, and nuclear wastes. Currently, a national debate is raging over the degree to which this should remain a last frontier free from major constraint or control.

A consequence of the continuation of the postwar convergence of urban and rural rates was that differences in urban-rural health rates, and health care, declined as an issue of interest to medical and social scientists. By the 1950s, urban death rates were still 20 to 25 percent higher than those in rural areas, but the common assumption was that rates were in the process of inevitable convergence (Bogue, 1959). The decade of the 1960s witnessed a resurgence of interest in urban topics, but this resurgence did not include concern with the spatial distribution of morbidity and mortality. Throughout the 1960s, the urban-rural differences that were found were assumed to reflect occupation, race, age, and sex composition. Higher urban rates of infectious diseases, lung cancer, and cirrhosis of the liver were treated as special cases (Fox et al., 1970).

PERSISTING DIFFERENCES

The dismissal of place of residence as a factor in health and health care, however, may have been premature. Today, any general statement that no subsantial rural-urban differences in health status exist requires three caveats. First, variations remain in the availability of health services. Second, some objective measures of health status continue to demonstrate clear locational differences. Third, nationwide surveys of persons' perceptions of their health and health care continue to show a consistent pattern of variation by size of place. Urbanites, small-towners, and rural dwellers differ significantly in how they evaluate their own health and health care. The following sections present recent data on these differences.

MEDICAL SERVICE VARIATIONS

Significant differences still persist between rural and urban areas in access to health resources. Data from the National Center for Health Statistics reveal that in 1977 the number of patient care physicians per 100,000 population ranged from 170 in metropolitan areas to 99 in semirural areas to 56 in rural areas. Rural-urban differences in the availability of Board-certified specialists were even greater (Schwartz et al., 1980). Moreover, physicians practicing in nonmetropolitan areas work longer hours and see considerably more patients per week than do those in metropolitan areas (American Medical Association, 1977; U.S. Department of Health and Human Services, 1981). This, in turn, acts to encourage new physicians to avoid such higher-work and lower-income areas.

In addition to the availability of health resources, recent data suggest rural-urban differences in the *use* of medical resources. One index of use is the number and percentage of persons who have not consulted a physician in two years. Data from the 1976-78 National Health Interview Surveys reveal that in metropolitan areas, 12 percent of total sample had no physician visits during the preceding two years, whereas, in the semirural areas, the percentage was 14, and over 15 percent in the rural areas.

The relative advantage enjoyed by city dwellers is qualitative as well as quantitative. For example, research hospitals

TABLE 2.1   Persons with No Physician Visits During Past Two Years,
According to Age, Race, and Location of Residence:
United States, 1976-78

| Race and Location of Residence | All Ages | Age ||||
		Under 17 Years	17-44 Years	45-64 Years	65 Years and Over
Total[1]	Percentage of Population				
All areas	12.9	10.4	13.1	15.6	13.5
Metropolitan	12.2	9.2	12.7	14.9	13.2
Nonmetropolitan	14.3	12.8	14.0	17.0	14.0
Semirural[2]	13.6	11.3	13.6	16.7	14.1
Rural[3]	15.1	14.7	14.6	17.4	13.9
White					
All areas	12.9	10.0	13.2	15.8	13.5
Metropolitan	12.3	8.9	12.8	15.2	13.1
Nonmetropolitan	14.1	12.1	14.0	17.0	14.1
Semirural[2]	13.5	10.9	13.6	16.6	14.0
Rural[3]	14.9	13.6	14.6	17.6	14.3
Black					
All areas	12.4	12.5	11.5	13.5	13.7
Metropolitan	11.2	10.4	10.8	12.5	14.2
Nonmetropolitan	16.1	18.5	14.3	16.9	12.6
Semirural[2]	15.3	14.9	14.0	18.9	15.9
Rural[3]	16.8	20.9	14.5	15.5	10.4

SOURCE: National Center for Health Statistics, data from the 1976-78 National Health Interview Surveys.
1. Includes all other races not shown separately.
2. Nonmetropolitan counties with at least one city of 10,000 population or more.
3. Nonmetropolitan counties with no city of 10,000 population or more.

and Board Certified medical specialists are overwhelmingly concentrated in metropolitan areas — city and suburban (Schwartz et al., 1980). The practice of medicine in a rural or small-town environment of necessity requires a commitment to the practice of a more "general" medicine. Currently there is considerable emphasis on this need for more generalist family practitioners — but prestige still goes to the city specialists.

Also, physicians, like other professionals, make locational decisions on the basis of lifestyle as well as professional criteria. Choosing a medical practice in a small town requires that not only the physician but also his or her spouse be agreeable to small-town life. Traditionally this has meant being willing to forego the advantages of being part of a medical community in a metropolitan area, and the other amenities of an upper-class urban lifestyle.

OBJECTIVE INDICATORS OF HEALTH

One of the most difficult aspects in discussing the question of urban versus rural health is establishing, with some degree of consensus, a definition of the concept. Despite its importance to our individual and collective lives, health continues to be an elusive notion that defies easy operationalization.

There are no universally accepted measures of national health status. The absence of such health status indicators stems, in part, from the limited "carrying capacity" of U.S. ideological and theoretical perspectives. American theoretical approaches to health have tended to treat it as a private, individual concern rather than a social concept.

Among health measures commonly used by the National Center for Health Statistics was infant mortality, life expectancy, and certain morbidity rates such as the incidence and prevalence of selected diseases such as tuberculosis. Unfortunately, systematic information on morbidity is not readily available, nor are available morbidity data nearly as reliable or complete as the data on mortality.

Life Expectancy

Life expectancy, while it is a hypothetical measure, is often used as an overall indicator of health levels in the population since it has the capability of summarizing the mortality rates of a given period of time in a convenient form. However, as a measure of health, it has the disadvantage of giving greater weight — some would argue weight out of proportion — to the relatively large number of deaths during infancy (Lerner and Anderson, 1963: 317).

At the turn of the century, the expectation of life at the time of birth in the United States was 47 years. As of 1982, life expectancy had increased to 74 years. Significant variations among subgroups in the population have always characterized life expectancy. The most notable differences are those found between males and females and whites and nonwhites. While the differences in life expectancy between whites and populations other than whites have declined by eight years since the turn of the century, differences between males and females have increased from two to almost eight years (U.S. Department of Health, Education and Welfare, 1977).

Age-Adjusted, Cause-Specific Mortality Rates

Geographic, rural-urban, or other differences in age-specific mortality rates are believed to reflect possible environmental problems or inadequate health care services or facilities. However, a major limitation of the use of mortality statistics for determining the health, or lack of health, of a specific place is the historical problem of misallocation of the place of death. Part of the difficulty is due to the historical placement of major medical institutions (hospitals, clinics, and so on) in large cities. For many years, residents of surrounding suburban and rural areas who died in these institutions were misallocated to the larger cities. On the other hand, institutions for the care of tuberculosis and certain other diseases were usually located in rural areas (Sydenstricker, 1933: 69-70).

Ascribing all deaths to the place where they happened to occur rather than to the place of residence was not corrected in the United States until the mid-1930s. This creates a problem in interpreting data from earlier periods.

Infant Mortality Rates

Since infant death is much more sensitive and susceptible to environment influences than mortality among adults, it has been widely used as an indicator of prevailing health conditions (U.S. Bureau of the Census, 1973: 410-411). While a useful measure of cross-national or cross-regional comparison, it has the obvious limitation of being relevant only by implication to the health and morbidity experiences of older persons.

Degree of urbanization has persisted as a relevant factor in infant mortality rate differentials, but unlike earlier centuries, urban areas now clearly have the advantage. These differences continued to be substantial in the 1970s. It should be noted that the greatest difference in infant mortality occurs not by region but by race. Findings show that suburban counties of large metropolitan areas (more than 1 million population) had the lowest rates — 12.3 deaths per 1,000 live births for white infants and 23.9 for black infants. The highest rates were found in counties with small cities, not adjacent to metropolitan areas — 15.1 per 1,000 births for white infants and 28.9 for black infants (U.S. Department of Health and Human Services, 1980).

Tuberculosis

The leading causes of death in the United States at the turn of the century were those diseases generally classified as communicable. In this category were the influenza/pneumonia group, tuberculosis, and gastritis. With a combined death rate of about 540 per 100,000 population, these diseases accounted for nearly one-third of all deaths in 1900 (Lerner and Anderson, 1963: 41). After the influenza/pneumonia group, which had a mortality rate at that time of 202 per 100,000, tuberculosis was the leading cause of death, with a mortality rate of 194.

The usefulness of rates of tuberculosis as a measure of health lies in the close relationship of the incidence (and prevalence) of the disease with living conditions, working environment, advances in disease detection, and other variables subject to social change and control. However, the very success in controlling tuberculosis has led to a dimunition of TB rates' effectiveness as a general measure of community health. Between 1900 and 1960, the death rate from tuberculosis declined by more than 97 percent to 5.5 per 100,000 (Lerner and Anderson, 1963: 41). By 1960, tuberculosis accounted for less than 1 percent of the total mortality in the United States (Lerner and Anderson, 1963: 50).

A study of mortality from tuberculosis in urban and rural communities of nine states for the period 1908-12 documented strikingly different rates for the respective areas, particularly for certain ages. Children under 15 years and adults between 30 and 50 years of age, living in urban areas, had 70 percent higher death rates than their rural counterparts (Sydenstricker, 1933: 71-73). The study concluded that higher mortality from tuberculosis among urban adults was accounted for largely by the deaths of males — supporting the notion that the disease was responding to urban environmental conditions, namely, poor working conditions.

While death due to tuberculosis is exceedingly rare today, the number of cases continues to constitute a health problem. In 1979, 27,669 cases of tuberculosis were reported to the Center for Disease Control for a *case* rate of 12.6 per 100,000. Moreover, tuberculosis continues to be an "urban disease," with the rate in large metropolitan areas more than twice that in small metropolitan and nonmetropolitan areas (U.S. Department of Health, Education and Welfare, 1977).

In the nation's 58 largest cities in 1973, the tuberculosis case rate was 25.7 per 100,000 population; in the 95 cities with 100,000 to 250,000 residents, the rate was 16.4; the case rate for all other areas (less than 100,000 population) was 11.3 (U.S. Department of Health, Education and Welfare, 1975). Thus, recent data indicate the higher urban rate pattern continues. There is, however, considerable city-to-city variation with 1979 rates, ranging from 54.5 per 100,000 in San Francisco to 3.4 in Omaha (Center for Disease Control, 1980).

Chronic Diseases

In the mid-1970s, it was estimated that approximately 30 million persons in the United States were, to some degree, affected by chronic diseases. However, the lack of complete and objective data on the incidence and prevalence of such diseases has limited their use as an indicator of health.

According to Fox, chronic disease morbidity (except for tuberculosis) usually is not included among diseases reported by states to the Center for Disease Control. Consequently, the only data on the occurrence of such diseases available on an official registration basis are on those that lead to death and are reported as such (Fox et al., 1970: 121-122). Semiofficial and incomplete data for selected chronic diseases exist in many state tumor registries or are collected by voluntary health agencies oriented to such specific diseases as multiple sclerosis or cystic fibrosis.

Overall, the data on chronic diseases appear to suggest that there are differences by area but that these differences are less clear-cut once variables of age, sex, race, and socioeconomic status are included. Suburbanites generally have the lowest chronic disease rates, probably reflecting their relatively advantaged socioeconomic status.

Oral Health

In our search for health statistics that might usefully measure variations by location, we also examined nontraditional measures. One of these was oral health. While oral health is not included among the more traditional indicators, there are a number of arguments that could be made for its inclusion. It is, for example, a fairly sensitive objective measure of a physical condition that closely corresponds to socioeconomic status.

Despite recent increases in the number of dentists, significant variations continue in the supply of dental manpower among states and regions and between metropolitan and non-metropolitan areas (U.S. Department of Health and Human Services, 1981). The American Dental Association in 1979 reported that the dentist-to-population ratio was 62 percent higher in metropolitan than in nonmetropolitan areas. The "average number of active civilian dentists per 100,000 population for all metropolitan areas in the United States was 60 compared with only 37 for nonmetropolitan areas" (U.S. Department of Health and Human Services, 1981: 69). The metro/nonmetro pattern is generally pervasive and can be found in all regions, divisions, and most states.

The dentist-to-population ratio is closely associated with city size. The ratio of dentists to population for metropolitan areas of one million or more residents is 22 percent higher than for metropolitan areas with fewer than one million residents (U.S. Department of Health and Human Services, 1981: 70). According to the American Dental Association, such rural-urban differences among dentists are mainly the result of residential preferences. Dentists, similar to other professionals, tend to prefer and choose urban locales with their social, recreational, and cultural activities.

DATA PROBLEMS

In our judgment, current attempts to carry out comparative research on health status — such as the effect of urbanization on health status — are often stymied by the plethora of different health measures. The present system places too heavy a burden upon individual health measures, such as infant mortality, which are then used as surrogates for a national health measure. We believe what is needed is a universally accepted national health measure, some index of health status.

Inevitably, there will be debates over which rates should be included in such an index. Infant mortality rates would, of course, be included as a component, as would selected age-specific mortality rates and morbidity rates for contagious diseases such as tuberculosis. However, whatever the variables and their weightings, the crucial advantage of such an index would be that it would have universal usage. Cross-sectional and longitudinal research and comparison would become far easier and more reliable.

Philosophical and clinical arguments that any composite "artificial" index that describes no specific measure of mortality or morbidity will be difficult to interpret and evaluate should not be given much weight. Similar arguments were made against now commonly used "artificial" measures such as the Dow Jones Stock Averages, the Consumer Price Index, and the Index of Economic Indicators — composite measures that have proven their empirical utility for assessing the nation's economic condition. Ironically, it is now possible routinely to make economic comparisons between New York, Boston, Chicago, Houston, and Los Angeles, but in spite of the overwhelming funds and efforts devoted to health care, we are still unable easily to make similarly routine health status comparisons.

That we cannot do so reflects a state of comparative underdevelopment in health status research. While health statistics were historically among the first statistics to be collected, the focus of health data has not changed dramatically from the turn of the century, when the emphasis was, of necessity, on the collection of complete data rather than on analysis. There are an abundance of data on health services and incidence of various pathologies, but most of the epidemiological data are published in a format that is not compatible with the units commonly used today to analyze social and economic patterns. Health data are still often published by states. Far more useful would be comparable data published by standard metropolitan statistical area (SMSA) or ZIP Code. Health data should routinely be collected using the same spatial units used in demographic, social, and economic research.[3]

PERCEPTIONS OF HEALTH CARE

While objective measures indicate that the central cities appear to have become at least as healthful as the countryside, the general population, both rural and urban, perceives cities as having a clear advantage, both in health care and overall health of the population. This perception is remarkable, since it is at variance not only with the historical patterns but also with the antiurban attitudes of the American populace (White and White, 1962; Palen, 1981).

Public images and ideology historically have characterized the city as sinful, dirty, crime-ridden, and unhealthy, and these beliefs still have force. Even today, a large proportion of Americans express a clear preference for rural or small-town residence. A 1977 HUD-sponsored study, done by Louis Harris, found that only 10 percent of the population preferred to live in a large city. The Harris study concluded:

> Americans are almost unanimous (82%) in rating large cities as the "worst place to raise children." Also, 62% rate large city public schools "worst"; 62% attribute the worst housing to the large city; 54% say the highest taxes are in the large city . . . a near unanimous 91% point to large cities having the highest crime rates [Harris, 1978: 5].

Contrasting sharply with these negative views are perceptions of what large cities offer in the way of health care. Large cities are viewed by most Americans (61%) as the best communities for health care. This view is shared by persons from rural areas, towns, and suburbs as well as city dwellers (Harris, 1979: 25). Given the populace's preference for rural and small-town residences, it is therefore very significant that general health and medical care are perceived as superior in cities. Overall, some 62 percent of city residents report their health services as "good" or "excellent," whereas the figure for suburbs is only 55 percent and that for small towns and rural areas 52 percent (Harris, 1978: 31).

PERCEPTIONS OF GENERAL HEALTH

Related to the question of preceived level of health care available is the question of self-assessed perceived level of physical health — that is, does the individual "feel" healthy or not? The distinction between the "objective" and "subjective" measures of health and the perceived aspects of health is connoted by the concepts of disease and illness. Disease refers to an "objective" phenomenon characterized by changes in the

TABLE 2.2   Reported Health and Community Size, 1973-78 and 1980:
"Would you say your own health, in general, is excellent,
good, fair, or poor?"

Community Size	Excellent		Good	Fair		Poor
	%		%	%		%
Large cities, suburbs, and surrounding areas, 25,000 and over	34.0	(74.9)	40.9	19.5	(25.1)	5.6
Medium-size cities and suburbs, 50,000-250,000	33.3	(76.3)	43.0	18.5	(23.8)	5.3
Small cities, 10,000-49,999	29.0	(67.3)	38.3	24.8	(32.6)	7.8
Towns and villages, 2,500-9,999	30.1	(70.8)	40.7	22.0	(29.2)	7.2
Rural areas	26.0	(67.5)	41.5	23.2	(32.5)	9.3
Total	31.7	(73.0)	41.3	20.5	(27.0)	6.5

NOTE: N = 10,572. Percentage of numbers in parentheses = 100.0.

proper functioning of the body as a biological organism (Coe, 1970). Illness, on the other hand, refers to a subjective phenomenon in which persons perceive themselves as not feeling well. Illness attaches primary importance to the individual's own perception of the state of his or her health.

Assessments of perceived health have been carried out annually for the past decade and are reported by the National Center for Health Statistics. Similar self-assessment data have been gathered as part of the National Opinion Research Center's General Social Survey. Both sources permit analysis of data by the size and metropolitan character of the community.

These data indicate that most people, regardless of location, report they are in good health. However, aggregate data from NORC's General Social Survey for the years 1973-80 reveal a crude pattern in which residents of small cities, towns, and rural areas report poorer health than do residents of large and medium-size cities. The most consistent finding from these annual surveys is that residents of medium-size cities appear to have the fewest complaints. One-fourth of the residents of medium-size cities and of metropolitan populations say their health is "fair to poor." However, one-third of those living in rural areas and small cities report "fair or poor" health (see Table 2.2). Nationally, there is a consensus that urban folk not only have better health facilities but also are healthier than rural folk.

RECAPITULATION

To sum up, the U.S. pattern of urban-rural differences in health care has undergone profound transformation over the past century. Cities that were once the graveyards of countrymen show remarkable advances in public health, sanitation, and provision of medical services. The various measures we use to evaluate health status indicate that since the turn of the century urban mortality and morbidity rates have been decreasing faster than their rural counterparts.

Today, the big-city advantage exists not only in the objective indices but also in people's perceptions. Cities are perceived not only to have better health care but also to have healthier residents. The perception of urban advantage is important insofar as it flies in the face of virtually all other measures (such as crime, school, housing, taxes) in which cities are perceived less favorably than rural or small-town areas.

NON-WESTERN PATTERNS

The previous pages have documented the transformation of the American city from health hazard to health provider.

Health professionals, health data, and the public at large all are remarkably in concert: Cities have the best health care. This pro-city attitude is held even by those segments of the population who would prefer living in a small town or rural environs.

Nor is this pro-city stance confined to the North American continent. Throughout the globe, city populations routinely receive better health care than do their rural counterparts. The consequence has contributed to an urban population explosion (Hauser, 1982). The contemporary non-Western pattern of urban health advantage differs radically from the Western urban experience, where, as noted earlier, cities were most often deadly to city dwellers, and cities grew — or even sustained themselves — only through heavy inmigration. By contrast, contemporary Third World cities' dramatic growth would continue to occur through natural increase, even if all inmigration were magically halted tomorrow.

The cities of Europe and North America grew rapidly during the nineteenth century, in spite of their high mortality rates. The cities of the developing world are growing even faster, because they have not only high rates of inmigration, but additionally high rates of natural increase. There are no signs of this pattern changing during the remainder of this century. Health care in the cities of developing nations may be sporadic and unevenly distributed, but it beats the rural alternative hands down.

Thus, in terms of adequacy of health care and mortality patterns, the cities of the Third World are *not* replicating the Western World pattern. This does not mean that all parts of developing world cities have access to adequate, or even marginal, sanitary and medical facilities and services. Characteristic of most Third World cities is dramatic intracity differences. This is also the case for some U.S. cities. New York's South Bronx and Chicago's West Side, for example, have infant mortality rates that can be characterized only as a national disgrace.

FUTURE NORTH AMERICAN POSSIBILITIES

Given current patterns, what changes might be expected in Canada and the United States between now and the turn of the century? One change is likely to be the end to the era of concentration of health resources. From the Civil War to the 1970s, America experienced a century of concentration of population, power, finance, and fashion in urbanized areas. Technological developments in transportation and communication both reinforced the importance of the core and extended its dominance to the end of the paved road and telephone lines. Throughout the first seventy years of this century, health and medical services mirrored other aspects of the society by concentrating the biggest and best in the metropolitan areas.

Now, however, communication and, to lesser degree, transportation technologies no longer favor centralization. For the first time since the War of 1812, the census data show greater rural than urban growth rates. Technology has overcome physical and space barriers. Today the question, "How far is it?" is almost universally answered with a temporal rather than spatial response, e.g., "About 35 minutes."

However, the degree to which technology becomes a substitute for propinquity is still not fully grasped. That Nissan (Datsun) is building its first automated American automobile plant not in the city, but in rural Tennessee, reflects far more than the desire to escape Detroit's wage scale. The designers and planners recognize that recent changes in transportation and communication technology have outmoded the need for factories to be located in urban centers.

Health care, however, still reflects the pattern of concentration in a few locations. All in this century, health and medical services and facilities have centralized and specialized. The family physician bringing health care directly to the home has been replaced by the hospital and medical center. Treatment, particularly specialized diagnosis and treatment, has increasingly become the prerogative of urban-based and urban-organized medical systems.

Attempts to decentralize, based in part on ideological belief in the virtues of simpler family practice medicine, have not been very successful. We hear persuasive arguments that what is wrong with contemporary health care is that there are too many specialists, too much technology, too much treatment, too much surgery, and too much medicine. Nonetheless, programs initiated in the 1960s to encourage young physicians to put down roots in underserviced rural areas have not fulfilled advocates' hopes (Greer, 1979). In spite of the emphasis on general practice, the best and brightest of health practitioners still chose to specialize in the city, with its research hospitals and extensive medical community.

However, the future may well see the current pattern of urban concentration of health services supplanted by a pattern of dispersal. Where subsidies and exhortations failed to break metropolitan areas near monopoly of medical services, a radical transformation and expansion of our communication capabilities may lead to greater dispersal.

As noted earlier, the data clearly indicate that most Americans would like to live well beyond the city (Harris, 1979). Contemporary communication technologies have the potential to allow this by breaking the cities' monopoly on skills and equipment. For example, advances in interactive communication technology can now make the most complex medical diagnostic procedures, and even treatment, available to the geographically remote areas of the country. A rural physician whose clinic is wired for two-way communication can simultaneously consult with specialists in diverse cities. A closed-circuit cable medical channel can also allow distant diagnosticians actually to see and speak with the patient. Technology has once again made it possible for the physician to come to the patient rather than requiring the patient to come to the physician.

Communication technology is overcoming spatial barriers to the point where a physician who prefers a rural lifestyle could

even, by means of a closed-circuit medical educational channel, attend medical lectures or hospital rounds.

It should be noted that the crucial questions today regarding
these possibilities are not technological but social. Technically
we can do it. However, social change has yet to be accepted.
Will these possible communication technologies actually be
utilized? It is one thing to discuss how spatial barriers can be
overcome; it is another to say that they will be overcome,
particularly in health care.

Arguing against adoption is the well-documented fact that
physicians are noted for their conservative social and political
orientations. This has historically been true of practitioners,
particularly those in small towns or small cities. Will nonurban
physicians be interested in wiring into high-tech medical access
systems?

Prediction is at best a dubious art, and we do not feel
comfortable predicting the adoption of specific technologies.
However, we are on somewhat safer ground when discussing
master trends in patterns of urban change — some of which
clearly will affect health care practice.

The first suggestion we would make is that in the United
States there is a continuing convergence in health of rural and
urban to the extent that by the turn of the century most differences are likely to have ceased to make a difference. Regardless of place of residence we are all increasingly urban in our
attitudes, outlooks, and orientations. Urbanism is the American way of life.

Our second suggestion is that health professionals in general — and physicians in particular — are among the most
urban-oriented professionals in the nation. As discussed earlier, physicians are overrepresented in urban places. And it is
those urban-oriented professionals who have led in the adoption of new technologies. Change is accepted not only in
strictly medical areas. Home computers, for example, are now
commonplace in the residences of medical and well as other
professional people. Significantly, this dramatic rate of adoption was not predicted. Even as recently as a decade ago

futurists saw home computers as a future rather than a prox-
imate change.

Based on the foregoing, we suggest that it will be exceptionally difficult for health professionals — whatever their personal
desires — to avoid the consequences of living in what is increasingly becoming a "wired" society. Nor will health professionals
be able to ignore the fact that after two centuries of concentration the nation is now dispersing its residences, industries, and
services. With this diminishing need for propinquity and the
resulting dispersal of population will come a dispersal of health
and medical services — both physically and through communication technology.

This in no way is meant to suggest that huge central-city
medical complexes will suddenly fade away; indeed,
downtown department stores did not close once widespread
auto ownership made outlying shopping more convenient.
Downtown medical complexes, like downtown department
stores, will continue to have a substantial, if diminished, clientele. However, the master trend for the remainder of this century will be dispersal rather than concentration. The general
practitioner's famous black bag will increasingly be supplanted
by the health practitioner's interactive computer's black box.

NOTES

1. There was no similarly compelling self-interest in providing poorer neighborhoods with equal levels of schooling, parks, street maintenance, street lighting, or
policing. These public services remained less equitably apportioned.

2. "Foreign stock" was the census term used to designate not only the foreign-
born but also those born in the U.S. of foreign parents. The assumption regarding the
latter was that, while they technically were citizens in practice, they were
"hyphenated-Americans" rather than "real" Americans.

3. Considerable research on measures of health status is being carried out by
Rand Corporation. The thrust of this research has been individual indicators rather
than aggregate measures. For more information, contact the National Center for
Health Statistics, 3700 East-West Highway, Hyattsville, Maryland.

REFERENCES

American Medical Association (1977) *Profile of Medical Practice*. Chicago: American Medical Association, Center for Health Services Research and Development.

Bogue, D. (1959) *The Population of the United States.* New York: Macmillan, 1959.

Center for Disease Control (1980) "Tuberculosis — United States, 1979." *Mortality and Morbidity Weekly Report,* June 27.

Coe, R. M. (1970) *Sociology of Medicine.* New York: Mc Graw-Hill.

Fox, J. P., C. E. Hall, and L. R. Elvaback (1970) *Epidemiology: Man and Disease.* New York: Macmillan.

George, D. (1964) *London Life in the Eighteenth Century.* New York: Harper Torch Books.

Glaab, C. N. (1963) *The American City,* Homewood, IL: Dorsey.

Glaab, C. N. and A. T. Brown (1976) *A History of Urban America.* New York: Macmillan.

Grant, M. (1924) *The Passing of the Great Race.* New York: Scribners.

Graunt, J. (1939) *Natural and Political Observations Made upon the Bills of Mortality.* Baltimore: Johns Hopkins University Press.

Greer, A. L. (1979) "Health care policy: disillusion and confusion," in J. Blair and David Nachmias (eds.) *Urban Policies in Transition.* Beverly Hills, CA: Sage.

Harris, L. (1978) *A Survey of Citizens' Views and Concerns About Urban Life.* Study P-2795, Department of Housing and Urban Development.

Harris, L. (1979) Survey reported in *Occasional Papers in Housing and Community Affairs,* No. 4, Office of Policy Development and Research, U.S. Department of Housing and Urban Development, July.

Hauser, P. et al. (1982) *Population and the Urban Future.* Albany: State University of New York Press.

Health Progress in the United States: A Report of Health Information Foundation, 1900-1960. Chicago: University of Chicago Press, 1963.

Lampard, E. F. (1973) "The urbanizing world," in H. J. Dyos and M. Wolff (eds.) *The Victorian City.* London: Routledge & Kegan Paul.

Lerner, M. and S. Anderson (1963) *Health Progress in the United States, 1900-1960: A Report of the Health Information Foundation.* Chicago: University of Chicago Press.

Lipscomb, A. A. and A. E. Bergh [eds.] (1904) *The Writings of Thomas Jefferson,* Vol. X. Washington, DC: Thomas Jefferson Memorial Association.

Palen, J. J. (1981) *The Urban World.* New York: Mc Graw-Hill.

Population Reference Bureau (1978) *Population Handbook,* Washington, DC: Author.

Riis, J. (1982) *The Children of the Poor,* New York: Scribners.

Rogers, E. S. (1960) *Human Ecology and Health.* New York: Macmillan.

Ross, E. A. (1914) *The Old World in the New.* New York: Century.

Schwartz, W. B., J. P. Newhouse, B. W. Bennett, and A. P. Williams (1980) "The changing geography of board-certified physicians." *New England Journal of Medicine* 303: 1032-1038.

Smith, S. (1911) *The City That Was.* New York: American Public Health Association.

Strong, J. (1891) *Our Country.* New York: Baker & Taylor.

Sydenstricker, E. (1933) *Health and Environment.* New York: Mc Graw-Hill.

U.S. Bureau of the Census (1973) *The Methods and Materials of Demography,* by Henry S. Shryock, Jacob S. Siegal, and Associates. Washington, DC: U.S. Government Printing Office.

U.S. Bureau of the Census (1975) *Historical Statistics of the United States: Colonial Times to 1970, Part I.* Washington, DC: U.S. Government Printing Office.

U.S. Department of Health and Human Services (1980) *Health: United States, 1980* Hyattsville, MD: National Center for Health Services Research.

U.S. Department of Health and Human Services (1981) *Health: United States, 1981*. Hyattsville, MD: National Center for Health Statistics.

U.S. Department of Health, Education and Welfare, Public Health Service (1975) *Health: United States, 1975*. Rockville, MD: National Center for Health Statistics.

U.S. Department of Health, Education and Welfare (1977) *Health: United States, 1976-77*. Rockville, MD: National Center for Health Services Research.

White, M. and L. White (1962) *The Intellectual versus the City*. Cambridge, MA: MIT Press.

3

The Reconfiguration of Urban
Hospital Care: 1937-1980

ALAN SAGER

□ THE CONFIGURATION OF URBAN HOSPITALS in the United States has been changing in ways that manifest this nation's unwillingness to finance equal access to needed health services. The consequences of these changes in hospitals' ownership patterns, locations, sizes, objectives, scopes of service, and costs have been to diminish our ability to finance and deliver equally accessible care.

In recent decades, public general hospitals, the traditional providers of last resort to the uninsured, have suffered massive bed reductions. Many now face financial calamity. Many smaller and less costly voluntary, nonprofit hospitals that have been heavily committed to serving minority and low-income patients have been obliged to close. Those that have survived, and larger voluntary teaching hospitals that share their commitment, are experiencing increasingly serious financial problems, in part because of past overbuilding by teaching hospitals.

AUTHOR'S NOTE: *Work on this chapter was supported in part by Grant No. 18-P-97038/1-03 from the Health Care Financing Administration and Contract No. HHS-100-80-0127 from the Office for Civil Rights, Department of Health and Human Services. The contents of this publication do not necessarily reflect the views or policies of the Department of Health and Human Services or any of its branches. The assistance of Deborah L. Dennis, Sylvia F. Pendleton, Ruth Daniels, Marianne Muscato, William J. McMullen, and a host of skilled research assistants is appreciated, as are comments by Sandra A. Bornstein and the editors.*

These and related hospitals reconfigurations reflect physician preferences, hospital strengths and interests, and prevailing distributions of purchasing power for health care. Resulting patterns of service are potentially very effective but are imbalanced in ways that overemphasize highly specialized curative care. The high cost of this care reduces our capacity and willingness as a society to afford equal access to even the most basic, necessary, and effective service.

As a result, we face the danger of providing more and more care to fewer and fewer well-insured citizens. In part because many less expensive hospitals serving lower-income and minority patients have been closing or relocating from cities, our poorest patients are being concentrated in the world's most expensive hospitals, or are being denied care except in emergencies. American physicians' and urban hospitals' search for the best health care has in important ways become both the enemy of the good — decent and effective affordable care for all — and the unintended ally of the worst — access to no care at all for growing proportions of our citizens.

This chapter's four aims are to describe the ways in which urban hospital care has been reshaped, analyze the forces responsible, assess the consequences of the changes noted, and offer a simple solution to the problems manifested and exacerbated by reconfiguration.

SCOPE AND METHODS

Certain cities, hospitals, and times were selected for study. Large and middle-size central cities were chosen for several reasons. They contain high proportions of minority[1] and lower-income residents whose access to health care was thought to be endangered. In 1980, the combined populations of 52 cities studied were 42.4 percent black or Hispanic, up from only 11.0 percent in 1940 (see Table 3.1). In part for this reason, disproportionate numbers of hospitals in these cities were

TABLE 3.1   Urban Residents, Hospitals, and Beds, 1937 and 1980[a]

Characteristic	1937/40		1980	
	Number	% of U.S. Total	Number	% of U.S. Total
Population (millions)	30.9	23.5%	36.3	16.0%
Black population (millions)	3.1	24.0%	10.5	39.7%
Hispanic population (millions)	0.3[b]	N/A	4.9	33.9%
% black	10.0%	–	28.9%	–
% Hispanic	1.0%	–	13.5%	–
% black or Hispanic	11.0%	–	42.4%	–
Number of hospitals[c]	637	12.6%	699	11.6%
Number of beds (thousands)[c]	146	30.6%	238	24.0%
Active MDs (U.S., thousands)	160[b]	–	435[b]	–
General beds/MD (U.S.)	2.22	–	2.05	–
MDs/1,000 population (U.S.)	1.21	–	1.94	–

SOURCES: 1937: *Directory of American and Canadian Hospitals* (1937); U.S. Bureau of the Census, *Historical Statistics of the United States*. 1940: U.S. Bureau of the Census, *Statistical Abstract of the United States, 1941*. 1980: American Hospital Association (1980); U.S. Bureau of the Census, *Statistical Abstract of the United States, 1981*.
a. 52 cities were studied.
b. Estimate.
c. Acute general and other special (of 50 or more beds for cities studied).

hypothesized to be vulnerable to financial or other stresses.[2] These cities' share of U.S. hospitals fell slightly between 1937 and 1980, but they retained almost one-fourth of all acute care beds. In 1980, the standard metropolitan statistical areas (SMSAs) of the cities studied held 43.8 percent of all U.S. residents and about 42.4 percent of all beds.

An initial focus of this investigation was on the 31 large central cities of U.S. SMSAs with one million or more residents in 1970. A representative group of 21 of the 94 central cities of SMSAs with 1970 populations ranging between one-quarter million and one million was added subsequently (Appendix A). Hospital behavior in areas of different sizes has been discussed elsewhere (Sager et al., 1982); findings on all cities' hospitals will be reported together here in the interest of brevity.

Short-term acute care general and "other special" hospitals of fifty or more beds were selected for study.[3] The changing configuration of long-term care facilities — psychiatric, tuberculosis, and other chronic disease institutions — was thought to merit separate attention owing to differences in financing and organizing care of these conditions. Furthermore, reconfigurations of long-term hospitals, with the exception of psychiatric facilities (Bassuk and Gerson, 1978; Clarke, 1979), have generally been less controversial than have changes in the acute care sector.

The short-term acute care hospitals studied comprise the vast majority of the institutions most Americans think of as "hospitals." There has been controversy over the meaning and appropriateness of these institutions' reconfiguration in recent years. Hospital closings have been applauded by some as an appropriate method of reducing undesirable overbedding and thereby cutting costs sensibly (Whalen, 1980; McClure, 1976; Rogatz, 1975) but have been decried by others as tending to increase both cost and maldistribution of services (Kleiman, 1980; Dallek, 1982; Sager, 1980). Similarly, some see the expansion of the teaching hospital sector as producing valuable improvements in quality, while others perceive an inappropriate emphasis on unequally affordable and spatially maldistributed specialty curative care, to the inevitable detriment of needed preventive services and accessible primary (physician) care.

The starting date of 1937 was selected for two reasons. First, it captures the results of the astonishingly rapid expansion of hospitals that took place between roughly 1890 and 1930. In this way, it provides a benchmark from which to begin measuring and analyzing the dramatic changes in urban hospital care following World War II. Second, excellent data on U.S. hospitals are available for 1937; and almost all large U.S. cities were tracted for the 1940 Census of Population, permitting descriptions of the residents of the neighborhoods around each hospital.

Hospital behavior has been studied during four periods: from 1937 to 1950, 1950 to 1960, 1960 to 1970, and 1970 to 1980.

Data used to describe hospitals' initial configuration and to understand why it has changed have been drawn from a variety of sources. Information on hospital size, ownership, scope of services, occupancy rates, and the like have been published in the 1937 *Directory of American and Canadian Hospitals* and subsequently in the *Guides* of the American Hospital Association (AHA). The tape of the AHA's 1969 Annual Survey of Hospitals provided most of the data on finances. Patient characteristics by race and payer status were provided by hospitals on the 1973 and 1981 Compliance Reports of the Office for Civil Rights, Department of Health and Human Services. The 1973 forms reported the number of private physicians on hospitals' staffs as well. The American Medical Association's (AMA) published directories provided data on interns, residents, and medical school affiliations of hospitals. Information on hospitals' demographic environments was compiled first by locating each institution on a large census tract map of its city and identifying an "area" including the tract containing the hospital and all contiguous tracts whose farther boundaries were roughly equidistant from the hospital. Changes over time in tract boundaries were noted. Published census data on race/ethnicity and income were then used to calculate the minority proportion of each hospital's area and its tract's income relative to the citywide median. These maps were also used to identify all other hospitals within one mile and within two miles of each institution. The number of beds, occupancy rates, costs, and similar characteristics of the hospitals near each institution were calculated. Finally, citywide and SMSA-wide population characteristics, bed-to-population ratios, and occupancy rates were compiled. Data on most hospital characteristics have been recorded for 1937/40, 1950, 1960/61, 1969/70/73, and 1980/81 for each of the 1,142 hospitals studied. Hospitals were tracked over time by means of the published AHA and AMA records, supplemented when necessary by communications with informed individuals in the various cities. Changes in ownership, mergers, closings, relocations, and construction of new institutions were noted. Mergers between hospitals located less than one-half mile apart were treated by subsequently consolidating the hospitals into one institution. None of the antecedent hospitals was considered to have closed. Mergers between hospitals located more than one-half mile apart were ignored for most purposes (unless they resulted in the closure of all beds operated at a site) because the principal concern has been with the amount, location, and type of care provided, not their legal accidents. Closing was defined as the discontinuation of acute inpatient care; this almost invariably meant a total cessation of service. Relocations were defined as moves of one-half mile or more. Almost all moves were considerably longer; most meant a departure from a city to one of its suburbs.

HOSPITAL CONFIGURATION, 1937 AND 1980

Changes in the number, ownership patterns, locations, sizes, objectives, scopes of service, and costs of acute care urban hospitals register the reconfiguration of these institutions. Between 1937 and 1980, there were net increases of 62 hospitals (9.7 percent) and 89,000 beds (60.4 percent) in the cities studied (see Table 3.2). These increases were not uniform. Several older cities experienced net losses of hospitals and beds, especially in recent decades. Hospitals' ownership patterns changed slightly. In 1937, about four-fifths of all hospitals were owned by voluntary, nonprofit organizations and the remainder was fairly evenly split between public (city, county, hospital district, or state university) and proprietary (for-profit) control. By 1980, the share of all hospitals operated under public and voluntary auspices declined slightly, though only the public hospitals actually dropped in number. After sketching major shifts within each of the three sectors, attention will be given to the extent and causes of public and voluntary nonprofit hospitals' reconfigurations.

There have been striking changes in the numbers of beds operated by the different types of hospitals between 1937 and

TABLE 3.2 Hospital Configuration, 1937 and 1980: Institutions and Beds

Characteristic	1937				1980			
	Public	Voluntary	Proprietary	Total	Public	Voluntary	Proprietary	Total
Number of hospitals[a]	70	505	62	637	63	527	109	699
% of hospitals	11.0%	79.3%	9.7%	100.0%	9.0%	75.4%	15.6%	100.0%
Number of beds	47,497	94,256	5,397	147,150	35,720	185,641	14,667	236,028
% of beds	32.3%	64.1%	3.7%	100.1%	15.1%	78.7%	6.2%	100.0%
Beds ($\bar{X}$)	678.5	186.6	87.0	231.0	567.0	352.3	134.6	337.7
Beds/1,000 residents	1.54	3.05	0.14	4.73	0.96	5.12	0.42	6.50

a. 50 or more beds.

61

1980. Almost one-fourth of all public beds were lost; public beds fell from one in three to one in seven; and the average public hospital declined in size by over 110 beds (16.4 percent). The number of public hospital beds per 1,000 residents fell by 37.7 percent, from 1.54 to 0.96 per 1,000.

Voluntary hospitals grew impressively over the entire period. Their share of beds increased from almost two-thirds to almost four-fifths of the total. The number of voluntary beds increased by 97.0 percent and the average voluntary hospital grew by 88.8 percent to over 350 beds. The ratio of voluntary beds per 1,000 urban residents rose by two-thirds to 5.12 per 1,000. Some of this reflected legitimate expansion to care for growing populations and to accept patients who were newly enfranchised financially by work-related health insurance or subsequently by Medicare for the elderly or Medicaid for some low-income citizens. But this writer believes, for reasons to be presented, that much of the increase in voluntary beds — and their growing concentration in fewer, larger, and more costly teaching hospitals — has been both unnecessary and the principal source of whatever degree of overbedding now prevails.

The proprietary sector also expanded rapidly, especially in recent decades in a few growing southern and western cities. Increases were recorded in the number, size, and total beds of for-profit hospitals, and in the number of such beds per 1,000 residents. Nonetheless, only 6.2 percent of all beds in the 52 cities were operated by proprietary institutions in 1980.

The net increase in proprietary beds obscures great instability. Rates of both closings and new hospital construction were highest in this sector (see Table 3.7). Proprietary beds' share in the large Northeast and Midwest cities fell from 3.8 percent in 1970 to 2.6 percent in 1980. At the same time, they rose from 6.8 percent to 15.9 percent of total beds in the large and growing cities of the South and West. Generally, smaller proprietary hospitals owned by their physicians have tended over time to convert to nonprofit status, usually for tax reasons, or to close. In newer areas and underbedded areas, physicians may found a hospital in order to be able to care properly for their patients, but proprietary chains are increasingly doing this for them. Occasionally, a chain may purchase a hospital in a declining area, "milk" its depreciation value for tax purposes, then close it.

CAUSES OF RECONFIGURATION

This chapter focuses on three types of reconfigurations: the decline in public hospitals' ability to serve the uninsured, the closing of smaller voluntary hospitals caring for disproportionate shares of lower-income and minority patients, and the growth of larger voluntary hospitals — especially those with teaching programs.

PUBLIC GENERAL HOSPITALS

These institutions have traditionally performed four functions: serving as providers of last resort for the poor, housing certain costly but indispensable services (such as trauma units), teaching, and research. In 1937, local governments paid most of the cost of their services, which were usually provided in large wards. Public hospitals in most cities probably provided most of the care received by patients unable to pay and lacking a private physician. Even in 1981, though much reduced in both number and proportion of beds, public hospitals served an estimated 36.3 percent of urban patients lacking Medicare, Medicaid, Blue Cross, or other third-party insurance (see Table 3.3). This is probably a conservative figure.

Large numbers of interns and residents continued their educations, at low salaries, in these facilities while playing central roles in caring for the uninsured. Indeed, most interns and residents in public hospitals were not paid in 1937; exceptions, such as those at Denver City Hospital, typically received as little as $30 per month. Almost all public hospitals were teaching facilities, typically with high ratios of house staffs of interns and residents per bed (see Table 3.4).

TABLE 3.3 Beds and Patients, by Ownership, Race, and Source of Payment, 1973 and 1981, 52 Cities

Characteristic	1973				1981			
	Public	Voluntary	Proprietary	Total	Public	Voluntary	Proprietary	Total
Beds[a]	46,173	174,442	11,279	231,894	35,720	185,939	15,160	236,819
Total census[a]	33,528	143,136	7,822	184,486	27,969	147,192	10,326	185,487
Medicaid census	9,861	18,640	885	29,387	5,920	18,532	1,045	25,497
Medicare census	3,892	30,400	2,160	36,452	3,769	37,733	2,728	44,230
Blue Cross/commercial census	6,768	70,759	3,259	80,786	4,504	68,569	4,701	77,774
"No charge" census	2,171	1,866	2	4,038	2,179	1,858	1	4,038
Minority census	16,084	32,730	1,751	50,566	11,375	39,612	2,480	53,466
Lack third-party payer	13,007	23,337	1,518	37,861	13,776	22,358	1,852	37,986
	Percentage of Total							
Beds	19.9%	75.2%	4.9%	100.0%	15.1%	78.5%	6.4%	100.0%
Total census	18.2%	77.6%	4.2%	100.0%	15.1%	79.3%	5.6%	100.0%
Medicaid census	33.6%	63.4%	3.0%	100.0%	23.2%	72.7%	4.1%	100.0%
Medicare census	10.7%	83.4%	5.9%	100.0%	8.5%	85.3%	6.2%	100.0%
Blue Cross/commercial census	8.4%	87.6%	4.0%	100.0%	5.8%	88.2%	6.0%	100.0%
"No charge" census	53.8%	46.2%	0.0%	100.0%	54.0%	46.0%	0.0%	100.0%
Minority census	31.8%	64.7%	3.5%	100.0%	21.3%	74.1%	4.6%	100.0%
Lack third-party payer	34.5%	61.5%	4.0%	100.0%	36.3%	58.8%	4.9%	100.0%

a. 1970 and 1980.

TABLE 3.4 Hospital Configuration, 1937/50 and 1980: Institutional and Neighborhood Characteristics

Characteristic	1937-50				1980			
	Public	Voluntary	Proprietary	Total	Public	Voluntary	Proprietary	Total
% teaching	91.4%	68.9%	18.8%	66.3%	96.9%	65.6%	5.8%	59.4%
House staff per 100 beds	6.2	4.0	1.1	3.9	11.2	6.0	0.5	5.7
Number of facilities[a]	16.4	13.5	7.5	13.3	33.6	24.3	14.0	23.7
Cost per admission[a]	$213	$154	$132	$158	$3579	$2462	$1769	$2459
Length of stay[a]	16.1	9.2	9.1	9.9	9.2	8.5	7.1	8.4
Occupancy rate[a]	79.9%	75.3%	71.5%	75.5%	77.8%	77.0%	64.3%	75.2%
Area % minority[b]	20.4%	14.1%	13.9%	14.7%	48.0%	41.5%	37.8%	41.6%
Tract income relative to city[c]	75.2%	87.2%	110.8%	87.7%	67.6%	87.7%	104.0%	86.7%

a. 1950 and 1980.
b. 1940 and 1980.
c. 1950 and 1970.

The quality of their care has been mixed. Some public hospitals have suffered from political patronage or bureaucratic rigidities from time to time. But many — perhaps most — continue to provide superb care for victims of trauma and other emergencies, and for many other patients requiring the specialized care of a large teaching hospital. At the same time, constrained funding and orientation to specialty care seem to have left some public hospitals without ability or inclination to provide adequate amounts of decent and effective routine inpatient or ambulatory (outpatient) services.

There are several explanations for the declining role of urban public general hospitals, two benign or at least acceptable, and one clearly undesirable. First, many chronic care patients have over time been discharged to nursing homes, recipients of more generous public funding beginning in the 1950s and accelerating with the passage of Medicaid in 1965. This change is reflected in public hospitals' declining length-of-stay, which in 1950 averaged 75 percent higher than that of voluntaries but fell to only 8 percent greater in 1980.

Second, the uninsured proportion of urban patients — those likeliest to depend on public hospitals — probably declined until about 1970 owing to growth of real incomes, increasing Blue Cross and commercial insurance coverage, and the steady upgrading of federal and state support through medical vendor payments, Kerr-Mills, Medicaid, and Medicare. Many former public hospital patients were therefore able to overcome financial barriers to admission to voluntary and proprietary institutions. Still, substantial shares of public hospitals' patients — 70 percent in 1973 and 68 percent in 1981 — have been covered by public or private insurance (see Table 3.5). They use public hospitals for reasons of habit, difficulty in gaining entry to other institutions, convenience, or preference.

Third, while local government revenues (the principal source of payment for the 32 percent of public hospital patients lacking insurance in 1981) have grown faster since 1950 than has

TABLE 3.5 Hospital Characteristics by Ownership/Affiliation, 1970/73 and 1980/81

Characteristic[a]	Public	Voluntary	Proprietary	All Hospitals
Beds, 1970 (sum)	46,173	174,442	11,279	231,894
Beds, 1980 (sum)	35,720	185,939	15,160	236,819
% change, 1970-80	−22.6%	+6.6%	+34.4%	+2.1%
Occupancy, 1970	75.9%	82.0%	75.2%	80.6%
Occupancy, 1980	77.8%	77.1%	64.3%	75.2%
% change, 1970-80	+3.7%	−6.1%	−10.7%	−6.0%
Length of stay, 1970	11.86	9.43	7.93	9.45
Length of stay, 1980	9.24	8.54	7.13	8.39
% change, 1970-80	−22.1%	−9.4%	−10.1%	−11.2%
Cost per admission, 1970	$1267.94	$772.23	$582.57	$791.03
Cost per admission, 1980	$3578.78	$2462.55	$1769.33	$2459.01
% change, 1970-80	+182.3%	+218.9%	+203.7%	+210.9%
Medicare % admissions, 1973	16%	23%	28%	23%
Medicare % admissions, 1981	21%	27%	28%	27%
Medicaid % admissions, 1973	28%	15%	13%	16%
Medicaid % admissions, 1981	23%	16%	14%	16%
BC/CC % admissions, 1973	26%	51%	46%	48%
BC/CC % admissions, 1981	24%	47%	48%	46%
Medicare/Medicaid/BC/CC % adm., 1973	70%	89%	87%	87%
Medicare/Medicaid/BC/CC % adm., 1981	68%	90%	90%	89%
Inpatient % minority, 1973	50%	28%	30%	30%
Inpatient % minority, 1981	55%	32%	31%	33%
"No charge," % admissions, 1973	9%	1%	0.03%	2%
"No charge," % admissions, 1981	6%	0.06%	0.0%	0.5%
Hospital transfers % admissions, 1973	3%	1%	0.38%	1%
Hospital transfers % admissions, 1981	7%	0.9%	0.5%	1.3%

a. BC = Blue Cross; CC = commercial carriers.

gross national product, hospital expenses per patient day have grown considerably faster (see Table 3.6). The resulting financial pressure on many cities' budgets has undoubtedly contributed to public hospital bed reductions and closings. Clearly, health care for the urban uninsured, financed principally and increasingly by the slowest-growing source of revenue, local governments' property tax, has been overmatched by society's fastest-growing cost, hospital care. In recent years, the concentration of low-income and uninsured patients — with problems that are typically more costly to treat (Horn, 1982) and whose numbers have been growing in response to Medicaid cuts and rising unemployment — in expensive public teaching hospitals has come to threaten these institutions' capacity to continue to perform their traditional function as hospitals of last resort (Dowling, 1982; Sager, 1982). Substitutes are distinctly lacking.

Relatively few public hospitals closed or relocated between 1937 and 1980 (see Table 3.7); most of these have done so since 1970, when financial pressures began to precipitate closings in St. Louis, Philadelphia, and New York City. Few public hospitals relocated great distances from the concentrations of lower-income patients they were built to serve. Two of those that did move far, in San Antonio and Hartford, were compelled to do so to be near a new state medical school. It is important to note that public beds have been cut principally by mothballing or demolishing large parts of surviving hospitals, not by closings. Many cities — including Boston, Washington, Cleveland, Louisville, Chicago, Kansas City, New Orleans, and San Francisco — lost between half and three-fourths of their public hospital beds. More recently, contract management of some public hospitals — by a medical school, voluntary teaching hospital, or private firm — seems to have been employed at least in part as a vehicle to limit local governments' financial contributions to their hospitals. A well-defined legal or political responsibility to serve as a provider of last resort can be obscured or avoided in this way. Sometimes, only a fixed annual payment for free care is appropriated by local government to a managed hospital. Though atypical today, this arrangement can be expected to occur more frequently.

TABLE 3.6 Hospital Costs and Municipal Resources, 1955-79

Year	Municipal Spending on Own Hospitals ($M)	% of Total Revenues	Hospital Costs per Patient Day	Gross National Product ($M)	Total Municipal Revenues ($M)	% Local	% from State	% from Federal Government
1955	$ 410	4.0%	$ 23.12	$ 328	$10,227	85.9%	12.1%	2.0%
1965	796	3.9	44.48	688	20,318	82.6	13.5	3.9
1975	1,986	3.3	151.00	1,529	59,744	67.1	21.8	11.0
1979	2,555	2.9	247.75[a]	2,369	86,864	69.7	17.8	12.5
% increase, 1955-79	523.2%	—	971.6%	622.3%	749.4%	479.5%[b]	1,151.8%[b]	5,264.9%[b]

SOURCES: U.S. Bureau of the Census, *Statistical Abstracts of the United States:* 1971, Table 633; 1975, Tables 127, 438; 1980, Tables 179, 482, 513. Bureau of the Census, *City Government and Finances in 1978-79*, Table 1.
a. Extrapolated from 1978: inflated by 11.6%. See *Health Care Financing Trends*, Winter 1981, Table B-2; inflation factor is for community hospitals.
b. In total revenues from each source.

Some responses of the public hospital sector to greater patient need but reduced financial resources during the 1970s have had positive aspects. Cost per admission grew slowest here in the 52 cities studied. Only here were beds cut and occupancy rates increased. In 1980, occupancy rates were slightly higher here than in the voluntary and for-profit sectors. By contrast, inefficient and expensive overbuilding of nonpublic beds continued during the 1970s and occupancy rates fell (see Table 3.5).

While this relatively efficient public hospital performance will not in itself be sufficient to save most public institutions from fiscal calamity during the 1980s — caused largely by increases in the uninsured population and by limited local financial resources — it does point to a perhaps unexpected compatibility between the aim of assuring equal access to needed services and that of controlling the rate of increase in spending on health care. Since only about two-thirds of public hospitals' patients were insured by Medicare, Medicaid, Blue Cross, or other third parties (compared with 90 percent in the voluntary and proprietary sectors), because public hospitals therefore faced greater difficulties in passing incurred costs through to insurers, and because most public hospitals served as institutions of last resort reluctant to control spending by turning away patients, these institutions seem on average to have learned to behave more efficiently. This suggests that efforts to assure more equal access to care can, under some circumstances, proceed with — and even help to inspire — sensible steps to control spending.

Regardless of the degree to which recent public hospital bed reductions have been appropriate responses to reduced need or unfortunate cuts forced by inadequate funds, it is clear that need will again rise in coming years while funding will be less adequate. Growing numbers of chronic care patients are finding it difficult to obtain admission to nursing homes. Medicaid and Medicare cuts will force acute care patients back into public hospitals. Even between 1973 and 1981, before the Reagan administration's budget cuts were implemented, the share of patients admitted to public hospitals who were transferred from other sectors more than doubled (Table 3.5). This increase will continue, as many voluntary hospitals serving the poor experience grave financial problems, and as many of them close. Ongoing cuts in real federal funding for community health centers will propel patients to seek ambulatory care at public hospitals.

In sum, rising costs will make it difficult for local public hospitals even to maintain their current levels of service to the uninsured. Expansion to meet growing needs will be constrained by lack of both operating funds and beds. Beds retired from service in the past can seldom be returned to use without extensive rehabilitation, necessitated by current life safety codes. Beds demolished in the past can be replaced either through very costly new construction or by leasing or purchasing some of the growing number of voluntary beds that cannot be filled owing to overbuilding or to declines in the number of adequately insured patients. Neither path will be taken unless funds for care of the uninsured are made available.

At the same time, public hospitals remain very costly. This is partly attributable to education of physicians and others, to research, and to the real cost of many expensive therapies that public hospitals often provide. To the degree that these activities are necessary, they should be paid for, though not disproportionately by local taxpayers. Regional and national functions should be funded accordingly. If this is not done, patients at public hospitals (and larger voluntary teaching institutions) will continue to receive care more costly than their medical needs indicate.

VOLUNTARY HOSPITALS

The absolute and proportional decline in public hospital care has been paralleled by increases in the size and share of the voluntary sector. The number of voluntary beds has grown in all decades and in almost all cities studied. (New York is one striking exception. But even in the eighteen larger Northeast and Midwest cities [Appendix A], which collectively lost 12.9

percent of their residents between 1970 and 1980, voluntary beds increased by 3.9 percent and voluntary beds per 1,000 residents by 19.1%.) This cannot be expected to continue.

Growth within the voluntary sector has not been evenly distributed either across or within cities. Across cities, beds were added at greater rates in the growing South and West. Within cities, larger and better established teaching hospitals and newer hospitals on the peripheries have been likelier to survive and grow. The focus here will be on intracity differences.

It is most useful and historically accurate to divide urban voluntary hospitals into two types. The first are the older, larger, more specialized, medical school-affiliated teaching hospitals. These were founded, often by the mid-nineteenth century, for many of the same reasons as were local public hospitals: principally to care for chronically or terminally ill and impoverished citizens. Initially, charges were rare. Those in ill health who could afford to pay for care preferred to be attended by their physician at home. In older voluntary institutions, private philanthropy played the role taken by local tax money in public hospitals.[4]

Several advances in medicine during the second half of the last century radically changed these early voluntary hospitals. Anesthesia, antiseptic surgery, radiology, and modern nursing permitted physicians to diagnose and treat more effectively. Hospitals gradually became sites of active therapy for all economic groups. Existing philanthropic hospitals frequently became allied with new or realigned medical schools that survived the Flexner Report (1910).[5] As care became more effective, both the number of patients who could be helped and cost per person grew. The necessary revenue could not be provided by philanthropists alone. The practice arose of charging wealthier and middle-income patients above the cost of care in their private or semiprivate rooms, and using the resulting profit to subsidize care for hospitals' lower-income patients in wards. In return, the latter were expected to continue to serve as teaching "material" for medical students, interns, and residents. Medical school faculty members and others of a city's better physicians cared for some wealthy and middle-income patients, and many who were poor, at voluntary teaching hospitals. Hospital interest in encouraging the cross-subsidization by income groups was raised a notch by each economic downturn (during which philanthropic contributions fell, seldom to regain their former share of hospital costs).

These teaching hospitals were inadequate to care for cities' growing populations or to serve as workshops for still-plentiful physicians beginning to seek admitting privileges during the last decades of the nineteenth century and the first decades of the twentieth. Some physicians therefore established small hospitals run for profit. More frequently, physicians lacking admitting privileges at long-established teaching hospitals became associated with businessmen or religious leaders who were outside their cities' established philanthropic circles to found small and middle-size (50- to 200-bed) neighborhood-centered hospitals. These institutions were often associated with a particular religious or ethnic group. They typically espoused the philanthropic goals of the longer-established teaching hospitals, but their boards and paying patients were usually less able to afford to underwrite care for large numbers of impoverished patients. These community hospitals tended to serve general practitioners and surgeons who maintained offices in nearby neighborhoods and whose patients paid the costs of their own care.

The distinction between teaching and community hospitals is not meant to be categorical. Rather, urban voluntary hospitals fell on a continuum in size, degree of specialization, orientation to teaching, closeness of medical school affiliation, and service to the poor. Variables reflecting these sorts of characteristics and goals have been useful in explaining hospital behavior during the years following World War II.

The reconfiguration of urban voluntary hospitals is hypothesized to have been influenced by interacting scientific, medical, institutional, and demographic forces. Voluntary hospitals have always competed for survival to some degree, but — in the absence of anything resembling a free market — the results of this competition cannot be endorsed a priori as legitimate or desirable.[6] To make this judgment, the causes and impacts of hospital reconfiguration must first be evaluated.

Given the uneven distribution of purchasing power for health care, successful hospitals are hypothesized to have been those institutions able to attract a sufficient number of well-insured patients and the physicians to admit and care for them. Larger and more specialized teaching hospitals are hypothesized to have had both greater ability and greater inclination to remain open in U.S. cities. Smaller institutions, relying more heavily on physicians in private practice, are hypothesized to have found it difficult to remain open — especially when located in minority or low-income neighborhoods — without working to acquire many of the characteristics of the teaching hospitals.[7]

As medical science has advanced during the years since World War II, hospital care has become more complex, costly, and potentially effective. In part for this reason, it has become more difficult in practice to identify appropriate roles for small and middle-size urban community hospitals. These institutions have to a great degree been delegitimized. Although almost half of all acute care hospitals in the country have fewer than 100 beds and almost three-fourths have fewer than 200 (American Hospital Association, 1980), a great share of the urban institutions of this size have come to be regarded by some as both inefficient and liable to provide low-quality care — either inherently or in practice (Berry, 1967; Spitzer, 1970; Health and Hospital Planning Council of Southern New York, 1974; Comprehensive Health Planning Council of Southeastern Michigan, 1980).

The view held here is that the perception of inefficiency in many smaller hospitals stems in part from their real financial distress, associated with forces having little to do with costs of providing care but, rather, with location near large populations of inadequately insured citizens. Similarly, the perception of lower quality of care may follow more from smaller hospitals' failure to practice at the frontiers of medicine, even though they may competently provide — or be capable of providing — needed routine and less specialized services. For this reason, patients with choice may prefer to enter teaching hospitals even for care of uncomplicated problems. In other instances, small hospitals may indeed provide care of inferior technical competence or effectiveness. But this is likely to be owing to the caliber of physicians attracted or relegated to these institutions, and the closing of the hospitals in which these physicians practice is not likely to upgrade either physician skills or the care received by their patients. This discussion will be resumed later in this chapter, under the heading of impacts of hospital reconfiguration; what should be noted now is that advances in medical science have tended to undermine the legitimacy of the smaller and middle-size urban voluntary hospital. Increasingly perceived as incapable of performing their functions effectively, economically, and reputably, many of these institutions sought to survive by adding beds and specialized services and teaching programs.

Physician pressures have thrust hospitals in this direction. Growing numbers of physicians have served residencies to become specialists; they have sought to recast the hospitals in which they practice to resemble those in which they trained. This has been especially true in large cities, where teaching hospitals to emulate are close at hand. In many of these cities, demographic change has accelerated the departure of neighborhood-based primary care physicians in private practice, leaving hospitals increasingly dependent on specialists to admit and care for patients (deVise and Dewey, 1972; Dorsey, 1969; Elesh and Schollaert, 1972).

At the same time, two powerful forces induced hospitals to try to establish or expand house staff positions in internship or residency (teaching) programs. The first was the increasing complexity of medical care and the consequent growth in the proportion of hospital patients who might require care from a physician at any time of the day or night. Second, as some hospitals lost private physicians owing to demographic change in their surrounding neighborhoods, and as physicians generally came into short supply relative to the growing numbers of beds, hospitals gradually lost their historic bargaining power over physicians. Physicians could refuse to work in outpatient clinics or to take on a share of nonpaying inpatients. They could urge the hospital to establish a residency program, which could provide night and weekend coverage for their own increasingly needy patients. They could also threaten to hospitalize their patients elsewhere if new equipment were not purchased or reconstruction not undertaken.

Hospitals therefore sought to expand or initiate teaching programs. In so doing, they were responding to their own needs to have house officers available to care for patients and to relieve private attending physicians (and probably also in hopes of upgrading care at the institution) rather than to the nation's need for certain types or numbers of trained physicians. It is worth noting in passing that much of the current oversupply of physicians in the United States, their lack in many developing countries, and their overspecialization, is attributable to this pernicious pattern. Smaller and less well-known hospitals that established marginal teaching programs, especially those not affiliated with medical schools, have for years faced difficulty in filling available slots. This can only become worse as the supply of foreign medical graduates is being choked off and as the number of students graduated by U.S. medical schools is cut in coming decades. The necessity of staffing hospitals with fully trained physicians on salary will become clear.

During the period studied, larger urban teaching hospitals have been hypothesized as more able and more willing to remain open and to grow. They already had the equipment and house staffs of interns and residents that privately practicing physicians were demanding. Especially when affiliated with medical schools, teaching hospitals were buffered from the financial and medical staff consequences of demographic change in the neighborhoods nearby because they and their physicians (many of whom were salaried by the hospital and therefore not dependent on neighborhood-based practices) drew patients from a relatively wide geographic area.

Traditionally committed to serving some patients unable to pay, for both philanthropic and teaching purposes, they preferred to remain even in lower-income or minority central-city neighborhoods. Sited here, they were also equally accessible from all parts of the metropolitan area and could therefore attract both suburbanizing well-insured patients and those with complicated problems from throughout the area. Serving a substantial proportion of well-insured patients provided much of the money with which to underwrite the cost of caring for the uninsured, as the purchasing power of the philanthropic dollar in health care shrank steadily.

Large teaching hospitals typically had better access to both philanthropic and public funds for plant and equipment repairs and additions than did smaller, less prestigious, and less politically powerful community institutions. This helped keep teaching hospitals attractive to physicians and well-insured patients with choice about site of work or care. It also provided transient profit to hospitals, because capital projects were reimbursed in a peculiar way, one that yielded cash surpluses in their early years. Most capital spending, unlike uncompensated care (that provided to patients unable to pay), was fully reimbursed by health insurers. Surplus cash in the years just following completion of capital projects could be used to help underwrite operating losses, including those resulting from provision of uncompensated care. Unfortunately, this practice came to parallel that of New York City during the 1970s: it is necessary to borrow greater amounts, more and more frequently, to finance regular operating losses. Still, this diversion of depreciation funds for a time supported many teaching hospitals' care for the uninsured.[8]

As the minority share of the populations of the 52 cities studied rose from 11.0 percent in 1940 to 42.4 percent in 1980, hospitals were disproportionately affected. Hospitals have for historic reasons been concentrated spatially in cities' older districts, the neighborhoods in which minority citizens usually reside. It is therefore not surprising that the average minority share of the populations of the areas around voluntary hospitals remained slightly above the citywide average, despite the tendency of hospitals in minority neighborhoods to close or relocate.

Demographic change is hypothesized to have tended to set in motion forces that have worked to undermine smaller community hospitals. Larger teaching hospitals have been able to respond better to these pressures, at least for a time. Community hospitals — especially if already weakened by failure to maintain their physical plants or to add or modernize equipment, and by the aging of the physicians on whom they relied — were usually less able to cope with the problems likely to attend demographic change.

Demographic change induced many neighborhood-based physicians in private, fee-for-service primary care[9] practice to follow their relocated patients to the suburbs. Newly trained physicians have not tended to set up practices in minority neighborhoods. Community hospitals located in minority, demographically changing, or low-income neighborhoods are hypothesized to have been likely to lack both sufficient physicians to admit and care for patients and enough adequately insured patients to sustain occupancy rates at levels high enough to pay the institution's bills. From this perspective, low occupancy rates may reflect the low purchasing power of patients willing to use a hospital as much as they do the needs of the patients able to use it. (Reciprocally, high occupancy rates may manifest overservice of well-insured patients.) Community hospitals located in changing neighborhoods typically lacked traditions of serving significant proportions of uninsured patients because they were founded as doctors' workshops. They also lacked the money with which to underwrite care for the uninsured — either philanthropic contributions or surpluses earned by overcharging well-insured patients (too few of the latter could be attracted over increasing distances to ordinary community hospitals) — and the house staff and medical school faculty to provide that care.

In response to demographic change, some smaller community hospitals, especially those that had been dependent on cohesive groups of physicians or patients, would have been likelier to relocate. Other institutions, lacking either the spur or the resources to relocate and unable to survive where they were sited, were forced to shut their doors. Frequently, the immediate cause of a closing might be a dramatically deteriorating physical plant or the death or retirement of a physician who had been admitting a substantial share of the paying patients, but it is hypothesized here that the more systematic forces just explored underlay these immediate causes.

The same forces that put smaller community hospitals — especially those located in minority, changing, or low-income neighborhoods — at a disadvantage bolstered larger teaching hospitals and peripheral community hospitals. Overbuilding of advantaged institutions, especially of medical school-affiliated teaching hospitals, is believed to have taken place. Reasonable explanations include oversupply of specialized physicians relative to the needs of well-insured patients, unnecessary care engendered in part by fee-for-service compensation of physicians or the need to fill existing beds, institutional vanity, financial irresponsibility induced in part by insurers' willingness to reimburse almost all costs incurred by hospitals in serving their insureds, and competition for well-insured patients or those with interesting illnesses.

Comparisons of hospitals closing with those remaining open and examinations of survivors generally confirm the foregoing hypotheses. Multivariate regression and logit analyses point strongly in this direction as well. Smaller, less specialized community hospitals and those located in minority neighborhoods were more likely to close during each of the decades examined. This pattern has become more pronounced over time, even as the number and proportion of hospitals closing or relocating has increased. Other data demonstrate increasing concentration of urban beds in teaching hospitals closely affiliated with medical schools. This is attributable to the closing/relocation of many smaller, unaffiliated hospitals, the association of many growing middle-size institutions with medical schools, and the addition of beds by those with long-standing affiliations.

During the forty-three years between 1937 and 1980, 124 voluntary hospitals of fifty or more beds closed, a number equal to 24.6 percent of those open in 1937; 78 hospitals (15.4 percent) relocated (Table 7). This closing or relocation of 40.0 percent of hospitals eliminated beds equal in number to over 30 percent of those open in 1937. Many new hospitals were built, disproportionately in growing cities. The number of voluntary hospitals peaked at 576 around 1970 and has dropped steadily since, reaching 527 in 1980. It continues to fall, as the rate of closing accelerates and as fewer new hospitals are constructed. More beds are being concentrated in fewer hospitals.

During each decade, hospitals that remained open had more beds at the beginning period than those that relocated. Those relocating were larger than those that closed. Hospitals remaining open during the 1970s, for example, averaged 330 beds in 1970; those relocating, 166; and those closing, 132. Similarly, hospitals remaining open during any period were significantly more likely to be teaching hospitals (see Table 3.8). In at least two of the three most recent decades, hospitals remaining open had significantly higher occupancy rates and ratios of house staff per bed, and were located in areas with smaller minority population shares. The utility of these and related characteristics in distinguishing hospitals that close is clear from separate analyses of hospital survival between 1950 and 1980 and between 1970 and 1980.

TABLE 3.7   Hospital Closings, Relocations, and New Construction, 1937-80[a]

Ownership	1937 Hospitals Open	1937-80 Closings	1937-80 Relocations	1937-80 New Construction	1980 Hospitals Open
Public	70	16	10	18	63
Voluntary	505	124	78	224	527
Proprietary	62	48	3	100	109
Total	637	188	91	341	699

a. Only hospitals of 50 or more beds at some time during this period are included. Rows do not sum owing to changes in ownership.

TABLE 3.8 Voluntary Hospitals Closing/Relocating, Contrasted to Those Remaining Open, All Cities, 1937-80

Characteristic[a]	1937-50			1950-60			1960-70			1970-80		
	C/R	Open	Sig.[b]	C/R	Open	Sig.[b]	C/R	Open	Sig.[b]	C/R	Open	Sig.[b]
Beds	88.6	195.1	.0001	133.0	199.4	.0006	162.8	271.8	.0001	166.1	329.9	.0001
% teaching	23.1%	68.4%	.0001	62.8%	79.2%	.0133	54.9%	73.4%	.0051	42.7%	70.8%	.0001
House staff/ 100 beds	3.0	4.1	.1152	4.0	6.2	.0050	5.2	6.2	.3507	3.5	6.7	.0003
Occupancy rate	–	–		73.3%	75.8%	.3953	69.3%	78.6%	.0006	73.9%	83.7%	.0001
Area % minority	19.9%	14.2%	.1985	35.9%	17.4%	.0001	48.1%	24.6%	.0001	46.7%	33.0%	.0011
Number of hospitals	25	469	–	43	466	–	51	497	–	82	484	–
% of hospitals	5.1%	94.9%	–	8.5%	91.5%	–	9.3%	90.7%	–	14.5%	85.5%	–

SOURCE: Data in this table appeared in A. Sager, "Why Urban Voluntary Hospitals Close," *Health Services Research*, Vol. 18, No. 3 (Fall 1983), and are reprinted by permission of the Hospital Research and Educational Trust.
a. Separate values are given for hospitals closing/relocating or remaining open during each period; values are for the year beginning the period. Most characteristics are expressed as means; others are proportions or means of proportions. Included are the hospitals with 50 or more beds at the beginning of a period in 51 cities (Minneapolis excluded).
b. Significance of difference between C/R and Open, t test.

Voluntary hospital behavior between 1950 and 1980 was examined in order to learn whether hospitals' characteristics at the beginning of a long period were useful in "explaining" or predicting statistically which hospitals would close during that entire period. The characteristics of interest were hospitals' number of beds, number of special facilities (a measure of degree of specialization), cost per admission, teaching status, number of house staff per bed, years at site (a crude measure of age of the hospital's physical plant and of the adjacent housing), area percentage minority, change in area percentage minority between 1940 and 1980, and change in total area population between 1940 and 1980. These characteristics were selected to capture the forces hypothesized to influence hospital survival. Data on all these variables could be compiled on 455 of 523 (87.0 percent) of all voluntary hospitals of fifty or more beds open in 1950. Of the 455, 70 (15.4 percent) closed by 1980. Figure 3.1 compares selected characteristics of hospitals closing between 1950 and 1980 and of those remaining open throughout the period. The expected differences were found.

Regression analysis of hospital behavior was even more revealing. Hospitals' 1950 area percentage minority was the most important characteristic distinguishing hospitals closing between 1950 and 1980 from those remaining open. It was followed by 1950 occupancy rate and number of facilities and by change in area percentage minority between 1940 and 1980. These factors should be seen as additive — hospitals in more heavily minority neighborhoods, with lower occupancy rates and fewer special services, and in neighborhoods whose minority populations were increasing were likeliest to close (see Table 3.9). The calculated R^2 of 22.2 percent indicates the proportion of the variability in hospital behavior explained statistically by these four characteristics. We have found that an R^2 of this size on this number of cases is usually associated with a predictive accuracy of 80 to 85 percent in distinguishing hospitals that close. This indicates that hospital closings have been consistently associated with the forces hypothesized earlier to be important.

More complete information has been available to analyze voluntary hospital closings between 1970 and 1980. As indi-

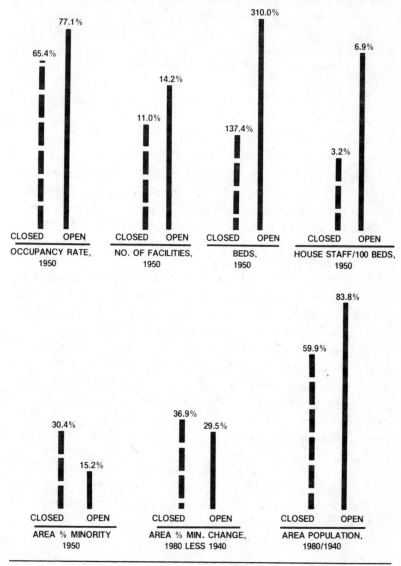

Figure 3.1   Comparison of Hospitals Remaining Open from 1950 to 1980 with
Hospitals Open in 1950 but Closed by 1980

TABLE 3.9   Regression Analysis of Hospitals Closing and
Remaining Open, 1950-80

Hospital Characteristic	Coefficient	Significance
Constant	1.6447	–
Area % minority (APM), 1950	−0.0083	.000002
Occupancy rate, 1950	1.1151	.000026
Number of facilities, 1950	0.0430	.000091
Change in APM, 1980 less 1940	−0.0047	.000184

NOTE: $N = 455$; $R^2 = 22.2\%$; significance = .000001.

cated in Table 3.10, hospitals closing during this decade were
smaller, had only about three-fifths the plant assets per bed, and
were slightly less expensive than hospitals remaining open.
They were much less likely to be religiously affiliated, suggesting either a weaker inclination to remain open or less of the
outside support needed to do so.

Hospitals closing were much less likely to operate teaching
programs; this was confirmed in their low ratios of house staff
per bed and higher ratios of private physicians per bed. Occupancy rates of hospitals closing were also lower.

Reliance on Medicaid-funded and minority patients at hospitals closing was about double that of surviving institutions.
Area percentage minority was greater, and only a small difference appeared in relative tract income. Unexpectedly, the
numbers of citywide beds per 1,000 residents were slightly
lower for the hospitals that closed. Overbedding did not seem
to cause closings. Citywide occupancy rates were about equal
for the two groups of hospitals. Those closing were sited near
greater numbers of other hospitals, though the occupancy
rates of the two groups of nearby institutions were similar. By
two measures, net revenues as a proportion of expenses and
deductions from revenue (partly for uncompensated care) as a
proportion of gross revenues, hospitals closings were in slightly
inferior condition.

Data on these and similar variables could be collected on
291 of the 487 (59.8 percent) voluntary hospitals of fifty or more
beds open in 1970. Thirty of the 291 (10.3 percent) closed by
1980. Regression analysis of hospital behavior during the 1970s
revealed eight characteristics significant at the level of .05 or

TABLE 3.10   Hospitals Closing and Remaining Open, 1970-80

Hospital Characteristic (1970)	Closing	Remaining Open
Number of beds	131.5	329.9
Plant assets/bed	$15,753	$26,959
Cost/admission	$736	$784
Religious affiliation	26.0%	46.6%
Teaching hospitals	33.9%	70.8%
House staff/100 beds	2.9	6.7
Private MDs/100 beds	123	99
Occupancy rate	70.8%	83.7%
Minority inpatients	49.8%	25.8%
Medicaid inpatients	27.0%	14.4%
Area % minority	51.2%	33.0%
Tract income/citywide	83.3%	87.0%
Citywide beds/1,000	6.78	6.89
Citywide occupancy	81.0%	81.5%
Beds within 2 miles	2,878	1,665
Occupancy rate within 2 miles	80.9%	81.3%
Net revenues/expenses	95.1%	96.5%
Deductions/gross revenues	12.1%	11.3%

SOURCE: Data in this table appeared in A. Sager, "Why Urban Voluntary Hospitals Close," *Health Services Research*, Vol. 18, No. 3 (Fall 1983), and are reprinted by permission of the Hospital Research and Educational Trust.

better (see Table 3.11). Most significantly, hospitals with greater inpatient minority shares were likelier to close. A hospital's area percentage minority or its inpatient percentage minority was a significant predictor of closings in both regression analyses for several reasons. In themselves, the two reflect social and economic inequalities, many of them products of discrimination. For smaller and less specialized hospitals, they indicate location in areas from which too few well-insured patients can be attracted, and in which too few physicians practice.

Hospitals with more beds nearby were likelier to close. This probably manifested some competition for paying patients and also greater likelihood of location in older and denser cities — and in older neighborhoods of those cities. As expected, hospitals with greater reliance on private physicians (denoted in Table 3.11 by those with "full" or "courtesy" admitting privileges) were also more likely to close, as were those in an inferior financial condition (measured by net revenues less expenditures as a proportion of net revenues). Occupancy rates of surviving hospitals were higher, reflecting the importance of attracting a sufficient number of well-insured patients. Hospitals remaining open also provided more outpatient visits per bed, a characteristic of large teaching hospitals.

This regression achieved an R^2 of 25.7 percent. Fully 88.6 percent of hospitals closing or remaining open were correctly categorized by the equation. Logit was used to analyze the same data, since it is theoretically preferable, though more expensive to use. Most of the variables significant in the regression were also significant in the logit. Perhaps more important, the logit predicted closings and survivals with 94.9 percent accuracy.

Regression analyses for both the thirty-year and ten-year periods indicate that smaller and less specialized urban voluntary community hospitals, especially those located in minority

TABLE 3.11   Hospital Closures: Significant Regression and Logit Variables, All Cities, 1970-80

| | Regression | | | Logit |
Hospital Characteristic	Coefficient	Significance	Sign*	Significance
Intercept	.7068	–	–	–
Inpatient % minority	–.1874	.0007	–	.0050
Beds within one mile (thousands)	–.0498	.0012	–	.0005
Full staff/bed	–.1001	.0137	–	.0050
Courtesy staff/bed	–.0684	.0140	–	.1200
(Net revenues-expenditures)/ net revenues	.3729	.0277		–
Occupancy	.3556	.0352		.0050
Outpatient visits/bed (thousands)	.2795	.0362		.1000
Construction in progress	.0564	.0377		.0250
House staff/bed	.4753	.0627		.1000

SOURCE: Data in this table appeared in A. Sager, "Why Urban Voluntary Hospitals Close," *Health Services Research*, Vol. 18, No. 3 (Fall 1983), and are reprinted by permission of the Hospital Research and Educational Trust.
NOTE: R^2 = 25.27%; n = 291; significant at .0001.
*Only negative logit signs are noted.

neighborhoods and serving high proportions of minority patients, have been much more likely to close.

Urban voluntary hospital care was also reshaped by changes made by surviving facilities. Ongoing hospitals that had been small in 1937 tended to have added beds by 1980. In the process they often started or expanded teaching programs and sought medical school affiliations for those programs. Surviving hospitals that had been large in 1937 also were likely to have added beds. A growing concentration of beds in voluntary hospitals with major affiliations with medical schools resulted from the observed patterns of closings and additions (see Table 3.12). In 1950, only 9.3 percent of the voluntary hospitals (with 18.6 percent of the beds) in the 52 cities had major medical school affiliations. By 1978, this rose to 32.1 percent of the hospitals (with 46.2 percent of the beds). The most rapid concentration of beds in hospitals with major medical school affiliations took place between 1969 and 1978, the general period during which access to urban health care ceased to become more equal — even as the share of gross national product devoted to hospital care continued to rise rapidly. The association between these two trends merits much closer examination than is possible here.

TABLE 3.12   Proportions of Voluntary Hospitals, and Their Beds, with Various Medical School Affiliations, 1950-78

Type of Affiliation		Year			
		1950	1961	1969	1978
None:	hospitals	82.3%	78.9%	68.0%	47.0%
	beds	69.9	68.3	51.6	29.2
Limited[a]:	hospitals	8.4	7.9	16.7	20.9
	beds	11.5	10.2	24.3	24.6
Major:	hospitals	9.3	13.3	15.3	32.1
	beds	18.6	21.5	24.1	46.2
Total[b]:	hospitals	549	573	588	526
	beds	115,868	142,764	173,284	184,783

a. In 1969 and 1978, includes "graduates only."
b. Numbers of hospitals and beds.

IMPACTS OF RECONFIGURATIONS

It seems clear that the observed pattern of reconfiguration is not moving us toward some desirable stable state of fewer, stronger, and more appropriately sized and located institutions able to serve their cities' patients. Rather, pubic hospital bed reductions and the closing of less costly voluntary hospitals (both serving high proportions of minority and Medicaid-funded patients) will oblige surviving hospitals — if they have room — to choose between denying care to displaced patients and admitting them, possibly lessening their own chances of remaining open.

Voluntary hospitals that survived the 1970s and were located within one mile of those that closed during that decade were (in 1970) already heavily committed to serving minority and Medicaid-funded patients. They were likely to be costly teaching hospitals, fully 44 percent more expensive per admission than those that closed (Table 3.13). In 1970, they were already in terrible financial condition. Possibly, many of these institutions managed to remain open until 1980 only by drawing on their capital — by diverting endowments and unfunded depreciation to cover operating losses.

Larger voluntary teaching hospitals lack the funds necessary to continue to care for large and growing numbers of uninsured or underreimbursed (as under Medicaid, in many states) patients. Philanthropic contributions seldom even cover a significant share of capital costs today. And commercial (non-Blue Cross) health insurers are in open revolt against the competitive disadvantage they face in financing teaching hospitals' traditional cross-subsidization from well-insured to underinsured patients (Health Insurance Association of America, 1982).

If patients displaced by closings receive alternative care, they — along with increasing proportions of low-income and elderly citizens insured under Medicaid and Medicare — will become concentrated in the world's most expensive hospitals. This helps contribute to public pressure to reduce eligibility, reimbursement rates, and scope of services covered under Medicare and Medicaid. Yet, the high apparent costs of urban

TABLE 3.13   Voluntary Hospitals Closing 1970-80, Compared with
Those Located within One Mile that Remained Open

Characteristic	Hospitals Closing 1970-80[a]	Hospitals Remaining Open	
		Within One Mile	All Hospitals
Beds, 1970	148.0	453.1	329.9
Occupancy, 1970	73.6%	80.0%	83.7%
Length of Stay, 1970	10.6	11.5	9.40[b]
Cost per admission, 1970	$638	$919	$784
Teaching hospital, 1970	43.2%	80.5%	70.8%
House staff/100 beds, 1969	6.0	11.2	6.7
Outpatient visits/bed, 1970	103.7	269.0	182.0
Admitted by private M.D., 1973	74.4%	58.3%	74.0%
Full physicians/100 beds, 1973	63.2	50.3	53
Courtesy physicians/100 beds, 1973	53.3	35.4	46
Private health insurance inpatients, 1973	34.5%	38.4%	51.0%[b]
Medicare inpatients, 1973	21.5%	29.9%	22.4%
Medicaid inpatients, 1973	29.1%	27.2%	14.4%
Minority inpatients, 1973	49.0%	35.4%	25.8%
Area % minority, 1970	48.9%	45.4%	33.0%
Deductions from revenues/gross revenues, 1970	15.0%	14.4%	11.3%
Deductions from revenues/bed, 1970	$4,042	$5,332	$3,802
Net revenues/expenses, 1970	95.2%	89.1%	96.5%
n	30	x̄30 groups	487

SOURCE: Data in this table appeared in A. Sager, "Why Urban Voluntary Hospitals Close," *Health Services Research*, Vol. 18, No. 3 (Fall 1983), and are reprinted by permission of the Hospital Research and Educational Trust.
a. Those with at least one hospital remaining open within one mile.
b. Estimate.

teaching hospital care probably exceed the real costs of services received by low-income and elderly patients in teaching hospitals. (This is probably one reason that teaching hospitals are willing to accept Medicaid patients.) A hospital's average room-and-board rates in part reflect teaching programs, some research, and considerable cross-subsidization from patients receiving routine care to those referred for more specialized service. It is the lower-income and elderly urban residents who are forced increasingly by the closing, relocation, or expansion of smaller and middle-sized hospitals to rely on costly teaching hospitals for routine care that could have been provided effectively in community hospitals. This second type of cross-subsidization, by diagnosis or severity of illness, increases financial pressures on the poor and elderly and on the programs paying for their hospital care.

Just as the disproportionate closing of lower-cost voluntary hospitals serving large shares of minority and Medicaid-funded patients probably works to reduce coverage under public programs — and thus financial access to care — so also the disproportionate closing of hospitals located in minority neighborhoods works to compromise spatial access to care. Although hospitals that closed were typically located in older downtown neighborhoods, near greater numbers of beds (on average), spatial access seems to have been harmed in several ways.

First, outpatient and emergency care at the closing hospital are lost. This particularly harms minority citizens, because they rely for physician care on outpatient departments two and one-half times as heavily as do nonminority citizens (National Center for Health Statistics, 1980). Second, remaining physicians in private practice in the neighborhood around the hospital, deprived of their organizational base, will be likelier to retire or relocate. Community health centers, an alternative source of ambulatory care, have been located disproportionately near hospitals that have closed. In fiscal years 1981 and 1982, fully 28.9 percent of these centers lost the special federal funding they needed to care for patients lacking insurance.

Access to inpatient care is reduced as well, particularly in vast districts of several cities from which most or all hospitals have closed or relocated. These include extensive sections of north St. Louis, south Atlanta, west Philadelphia, and other cities. In areas like these, few organizations with stakes in promoting or providing access to ambulatory or inpatient care remain. The diminished access resulting from many successive hospital closings in minority neighborhoods, in combination with the precarious financial conditions of surviving institutions serving high proportions of Medicaid-funded and minority citizens, indicates a serious problem.

Citywide data suggest that public and voluntary hospital reconfigurations have taken place at the same time that access to care for minority and Medicaid-funded patients has diminished. Between 1970/73 and 1980/81, the minority share of the 52 cities' populations rose from 35.0 to 42.4 percent, but the minority share of hospital inpatient censuses rose only from 27.4 to 28.8 percent. Census per 1,000 minority citizens fell 7.2 percent, from 3.74 to 3.47, and Medicaid census per 1,000 population fell by 7.9 percent (Table 3.14). While not age- and sex-adjusted, these changes do suggest a need for closer study. Perhaps as significant, the share of the total census financed under Medicaid fell by 13.5 percent between 1973 and 1981. This meant a loss of about 3,900 Medicaid-funded patients daily. A maximum of about 2,400 of them could have been displaced by the closing or relocating of the 83 voluntary hospitals that shut their doors during the 1970s — and many of these patients doubtless secured alternative care. Therefore, surviving hospitals had to have served fewer Medicaid-funded patients. Declining spatial access, state cuts in Medicaid eligibility, and perhaps some hospitals' reluctance to accept Medicaid patients contributed to this frightening drop in care provided to those sponsored by Medicaid. It must be noted that all of these

TABLE 3.14   Care of Minority and Medicaid-Funded Patients, 1970/73 and 1980/81, 52 Cities

Characteristic	1970/73	1980/81	% Change, 1970/73-1980/81
Minority population (thousands)	13,504	15,400	+14.0%
Total census	184,486	185,487	+0.5
Minority census[a]	50,566	53,466	+5.7
Medicaid census[a]	29,387	25,497	−13.2
Minority % total population	35.0%	42.4%	+21.1
Minority % total census	27.4%	28.8%	+5.1
Medicaid % total census	15.9%	13.8%	−13.5
Total census/1,000 population[b]	4.78	5.06	+5.9
Minority census/1,000 population[c]	3.74	3.47	−7.2
Medicaid census/1,000 population[b]	0.76	0.70	−7.9

a. Reconstructed from data on 1973 and 1981 Office for Civil Rights Hospital Compliance reports and AHA *Guides*.
b. Demoninator is total population, 52 cities.
c. Denominator is total minority population.

indicators of reduced access to care antedated the program cuts of the Reagan administration.

The potential effectiveness of the 52 cities' surviving hospitals — measured by the types of useful care they could competently provide — was probably greater in 1980 than at any earlier time. Regrettably, our society's willingness to finance equal access to this care is declining. Furthermore, surviving hospitals' configuration is not well matched with urban residents' routine needs for care. As urban Americans are forced by the closing, relocation, or transformation of small and middle-size hospitals to rely increasingly on large and specialized teaching hospitals for their inpatient and ambulatory care, they may be exposed to unnecessarily complex, costly, and uncoordinated care. As patients with routine problems are concentrated in teaching hospitals, they become available to absorb some of the transferred costs of esoteric and expensive specialized treatments provided to patients referred (often from a distance) to receive them. As a result of this second type of cross-subsidization, doctors, hospital administrators, and payers may not perceive the full costs of specialized care. Subsidization of specialized care may persuade us, as a society, to pay for more of it than we would if we knew its true cost. This contributes to increasing spending on highly specialized care, particularly because esoteric care on the frontiers of medicine is central to the interests of many urban physicians and hospital administrators. Such spending may or may not be appropriate, but it should be evaluated on its merits and in comparison to competing objectives, such as insuring universal financial access to all routine physician and hospital care.

In Boston, which experiences the highest hospital cost per admission in the nation — largely because it has gone furthest toward concentrating care in medical school-affiliated teaching hospitals — the Harvard Community Health Plan, the largest prepaid group practice in the state, has acquired control of a 100-bed hospital in which to serve inexpensively those of its members who require only routine inpatient care (Health Planning Council for Greater Boston, 1978). This illustrates the possibility of providing competent and effective basic hospital care in small institutions, while reducing cross-subsidization by severity of illness. The perception that good care for any problem is possible only in the best and most expensive hospitals must be combatted. The best way to do so is to upgrade the effectiveness and technical competence of care in smaller hospitals — not to close them.

More money will be needed to upgrade care in many smaller hospitals and to finance service in hospitals heavily committed to serving uninsured or underinsured patients. The well-to-do have always helped to pay for care of the less well-off in this country. Traditional arrangements for doing this are collapsing. Intrahospital cross-subsidization from some patients to others is no longer sufficient; it is also under strong attack by the insurers of those who are charged above cost. Some hospitals located in more affluent areas have freely channeled surplus revenues to other institutions; this interhospital cross subsidization also is insufficient. Direct public action is required to mobilize the sums necessary to pay for hospital care for all. At the same time, public attention to the configuration of care will be desirable, in the interest of ensuring affordable and well-placed services.

A SIMPLE SOLUTION

Urban hospital reconfiguration has manifested and exacerbated problems that can be solved by legislating health insurance coverage for all Americans. This proposal was considered seriously as recently as the early 1970s but was deferred until the rate of increase in costs was controlled. In retrospect, this apparently sensible decision was a mistake. Real cost control may be possible only when a concomitant commitment is made to universal access. Costs certainly have not been controlled in the years since the early 1970s. The *share* of gross national product devoted to hospital care alone increased by 44 percent between 1970 and 1981, even as access to that care probably became less equal. (During the prior three decades, by contrast, rising shares of gross national product purchased more equal access to hospital care.)

Two examples can be found of the ways in which universal access and cost control are allies, not enemies. One, noted earlier, is in the apparently growing efficiency of the public hospital sector, that closest to assuring unrestricted access within fixed budgets. The other is in the state of Maryland, where stringent budget review and cost ceilings have been tied to financial security for all needed community and teaching hospitals, in part through assumption of uncompensated care costs (by distributing them among the insurers of patients using the hospital). This disguised tax closely resembles universal health insurance; it was agreed to only after strict and effective controls over hospital spending were instituted (Robinson, 1982).

Universal financial access to care should be accompanied by fixed budgets for organized and accountable providers. Public payers should require universal access in exchange for stable financing of hospitals; hospitals should become increasingly willing to accept fixed budgets and accountability for certain patients in exchange for stable and adequate financing. Hospitals or health maintenance organizations responsible for defined populations or competing for members could receive these budgets.

These responsible providers would have clear incentives to save money by working to eliminate care that is ineffective, unnecessary, or incompetently provided. The $300 billion we now spend on health care — 10 percent of gross national product — is almost certainly sufficient to provide virtually all effective care to all citizens who need that care. Once all patients have financial access to care, responsible providers will need to work to reallocate much of that $300 billion to raise the health care system's floor — the worst it can do — and for a time to pay less attention to the ceiling — the best that can be done (Rosenthal, 1976; Cochrane, 1972). This will benefit growing shares of patients by easing worries about the affordability, accessibility, competence, or effectiveness of care. It will reverse the trend toward development of potentially more effective but inherently unequally affordable therapies — of which the artificial heart is only the most tragic of false hopes.

Responsible providers subject to fixed budgets will need to assert a measure of control over physicians. This will become easier for them to do over time, as physicians increasingly declare their preference — or at least their willingness — to work for salaries. Furthermore, the power relations between doctors and hospitals have been reverting toward those that obtained decades ago, as the supply of physicians grows while the number of hospital beds levels off (Table 3.1). The desirability of allowing physicians to escape hospitals' influence by establishing small for-profit facilities such as "surgicenters" should be carefully weighed. Responsible providers will be supported politically by the growing numbers of patients who would be forced to pay greater sums for health care — or lose access — if costs are not controlled.

Universal access to effective care, however financed, can save urban hospitals that serve high proportions of low-income, minority, and uninsured patients. In the face of recent Medicaid and Medicare cuts, surviving hospitals have even greater reason to become more effective advocates for their patients (Special Committee, 1982). With universal access will come new pressures on administrators, physicians, and researchers to identify, provide, and develop effective and affordable therapies and competent providers. Intelligent responses to these pressures will help our health care system to meet realistic public expectations for care during the decades to come. Once all patients are guaranteed financial and spatial access to care, we may have greater confidence that well-planned competition among responsible organizations will help ration inevitably limited care by efficiently providing the most effective services to the patients who need them. One way to build a foundation for appropriately planned competition will be to work to retain and restore a balanced configuration of community and teaching hospitals.

(Appendix A appears on page 96)

APPENDIX A   Study Cities

I. Larger Cities: Central cities of SMSAs of 1,000,000 or more in 1970

   A. Northeast and Midwest (HHS Regions I, II, III, V, and VII)

Baltimore	Kansas City, MO
Boston	Milwaukee
Buffalo	Minneapolis[a]
Chicago	New York[b]
Cincinnati	Newark
Cleveland	Philadelphia
Columbus, OH	Pittsburgh
Detroit	St. Louis
Indianapolis	Washington, DC

   B. South and West (HHS Regions IV, VI, VIII, IX, and X)

Atlanta	Portland, OR
Dallas	San Diego
Denver	San Francisco
Houston	San Jose
Los Angeles	Seattle
Miami	Tampa
New Orleans	

II. Smaller Cities: Central cities of SMSAs of 250,000-1,000,000 in 1970

   A. Northeast and Midwest

Bridgeport, CT	Patterson, NJ
Fort Wayne, IN	Peoria, IL
Gary, IN	Toledo, OH
Hartford, CT	Trenton, NJ
Jersey City, NJ	Wilmington, DE

   B. South and West

Austin, TX	Norfolk, VA
Beaumont, TX	Phoenix, AZ
Charlotte, NC	San Antonio, TX
Columbia, SC	Tucson, AZ
Louisville, KY	Wichita, KS
Memphis, TN	

---

a. Included in discussions of reconfigurations and closings, 1937-80; excluded from subsequent descriptions and multivariate analyses.
b. Five boroughs treated individually.

NOTES

1. Blacks and Hispanics were defined as "minority" in the present study. This was done principally because census tract data were available on these two groups (for the U.S. censuses of 1940 through 1980), because blacks and Hispanics comprise the great bulk of the central-city minority populations of interest, and for ease in calculating the necessary data on the areas containing the study's 1,142 hospitals.

2. See Sager et al. (1982: Chapter 2) for more detailed information.

3. These were almost all urban hospitals with average lengths-of-stay below thirty days, including general medical and surgical, pediatric, obstetric, and eye and ear hospitals.

4. This brief history draws in part on Belknap and Steinle (1963), Corwin (1946), Davis (1927), and Rosen (1963).

5. Flexner (1910) accelerated the demise of smaller for-profit medical schools and the growth of four-year university-affiliated, nonprofit institutions requiring prior undergraduate degrees. These resulted in fewer, better trained, and better paid physicians.

6. As Elliott Sclar has pointed out, the United States has for decades had one of the most competitive health care systems in the industrialized world.

7. See Johnson (1967) for an early and lucid statement of this position.

8. At the same time, the lure of depreciation revenues probably spurred unnecessary overbuilding. Appreciation of these issues has been broadened by Susanne Batko and Thaine Allison, Jr.

9. Primary care physicians include general practitioners, internists, pediatricians, and obstetrician-gynecologists. Boundaries are not firm: Specialists provide some primary care, just as some primary practitioners deliver specialized care.

REFERENCES

American Hospital Association (1980) *Hospital Statistics.* Chicago: Author.

Bassuk, E.L. and S. Gerson (1978) "Deinstitutionalization and mental health services." *Scientific American* 238, 2: 46-53.

Belknap, I. and J. G. Steinle (1963) *The Community and Its Hospitals.* Syracuse, NY: Syracuse University Press.

Berry, R. E. (1967) "Returns to scale in the production of hospital services." *Health Services Research* 2, 2: 123-139.

Clarke, G.J. (1979) "In defense of deinstitutionalization." *Health and Society* 57, 4: 461-479.

Cochrane, A.L. (1972) *Effectiveness and Efficiency.* London: Nuffield Provincial Hospitals Trust.

Comprehensive Health Planning Council of Southeastern Michigan (1980) *Plan for the Reduction of Excess Hospital Capacity in Southeastern Michigan.* Detroit: Author.

Corwin, E.H.L. (1946) *The American Hospital.* New York: Commonwealth Fund.

Dallek, G. (1982) "The continuing plight of public hospitals." *Clearinghouse Review* 16, 2: 97-101.

Davis, M.M. (1927) *Clinics, Hospitals, and Health Centers.* New York: Harpers.

de Vise, P. and D. Dewey (1972) *More Money, More Doctors, Less Care: Metropolitan Chicago's Changing Distribution of Physicians, Hospitals, and Population: 1950-1970.* Chicago: Regional Hospital Study.

Directory of American and Canadian Hospitals (1937) Chicago: Physicians Record Company.

Dorsey, J. L. (1969) "Physician distribution in Boston and Brookline." *Medical Care* 7, 6: 429-440.

Dowling, H. F. (1982) *City Hospitals: the Undercare of the Underprivileged.* Cambridge, MA: Harvard University Press.

Elesh, D. and P. Schollaert (1972) "Race and urban medicine: factors affecting the distribution of physicians in Chicago." *Journal of Health and Social Behavior* 13, 3: 236-250.

Flexner, A. (1910) *Medical Education in the United States and Canada.* New York: Carnegie Foundation for the Advancement of Teaching.

Health and Hospital Planning Council of Southern New York (1974) *General Hospitals and Related Services in Brooklyn.* New York: Author.

Health Insurance Association of America (1982) *Cost Shifting: The Hidden Tax on Health Care.* Washington: Author.

Health Planning Council for Greater Boston (1978) *Harvard Community Health Plan Merger with Parker Hill Hospital: Reviewer's Guide.* Boston: Author.

Horn, S. D. (1982) Personal communication, Center for Hospital Finance and Management, Johns Hopkins Medical Institution.

Johnson, R. L. (1967) "Urban hospitals face three choices: move, grow, or change." *Modern Hospital* 115, 11: 92-97.

Kleiman, M. (1980) "The unkindest cut of all." *Health/PAC Bulletin* 12, 1: 15-16, 25-38.

McClure, W. (1976) *Reducing Hospital Bed Capacity.* Minneapolis: Inter-Study.

National Center for Health Statistics (1980) *The Nation's Use of Health Resources: 1979.* Washington: U.S. Government Printing Office.

Robinson, M. L. (1982) "Marylanders call rate setting programs 'wave of the future.'" *Modern Healthcare* 2, 1: 36-38.

Rogatz, P. (1975) "The case for closing hospitals." *Medical Economics* 52 (February 3): 150-157.

Rosen, G. (1963) "The hospital: historical sociology of a community institution," pp. 1-36 in E. Freidson (ed.) *The Hospital in Modern Society.* New York: Macmillan.

Rosenthal, G. (1976) "Setting the floor: a missing ingredient in an effective health policy." *Journal of Health Politics, Policy and Law* 1, 1: 2-4.

Sager, A. (1980) "Urban hospital closings in the face of racial change." *Health Law Project Library Bulletin* 5, 6: 169-181.

Sager, A. (1982) "Causes and consequences of the public general hospital crisis," pp. 530-544 in Subcommittee on Health and the Environment, U.S. House of Representatives, *Extension of Health Planning Program.* Serial 97-126. Washington: U.S. Government Printing Office.

Sager, A., D. Dennis, and S. Pendleton (1982) *The Closure of Hospitals That Serve the Poor.* Waltham, MA: Heller Graduate School, Brandeis University.

Special Committee on Federal Funding of Mental Health and Other Health Services (1982) *Health Care: What Happens to People When Government Cuts Back.* Chicago: American Hospital Association.

Spitzer, W. V. (1970) "The small general hospital: problems and solutions." *Milbank Memorial Fund Quarterly* 48, 4: 413-447.

Whalen, R. P. (1980) "State concerns and role in hospital closure," pp. 43-48 in *Proceedings of the Health Policy Forum on Hospital Closures in New York City,* No. 2. New York: United Hospital Fund.

4

Mortality, Morbidity, and the Inverse Care Law

JOHN B. McKINLAY
SONJA M. McKINLAY
SUSAN JENNINGS
KAREN GRANT

☐ THE OBSERVATION THAT SUBSTANTIAL differences remain in the experience of morbidity and mortality by various social groups, despite rising living standards and demographic changes, is neither new nor startling. This notion, best summarized as the "inverse care law" — that the availability of good medical care tends to vary inversely with the need for it in the population served (Hart, 1971) — served as a description of a less-than-optimal outcome of the British health policy. That policy, similar in many ways to health and social policies in this country, was designed to alleviate social disparities, but the goal of equality of health outcomes has yet to be realized. The following discussion examines the applicability of the inverse care law in America and considers as well the limitations of medicine in contributing significantly to improvements in health. Although righting the wrongs of the inverse care law is a worthwhile mandate of health policy, it is argued that we must look beyond the issue of equality of access to the more important issue of how to allocate rationally scarce resources for the good of all.

Habakkuk (1953), an economic historian, was probably the first seriously to challenge the prevailing view that the modern increase in population was due to a fall in the death rate affected by medical interventions. His view was that it resulted from an increase in the birthrate, which was associated with social, economic, and industrial changes occurring during the eighteenth century.

McKeown has pursued the argument more consistently and with greater effect than any other researcher. Employing the data and techniques of historical demography, McKeown (a physician by training) provided a convincing analysis of the major reasons for the decline of mortality in England and Wales during the eighteenth, nineteenth, and twentieth centuries (McKeown and Record, 1955, 1962; McKeown et al., 1975). For the eighteenth century, he concludes that the decline was largely attributable to improvements in the environment. His findings for the nineteenth century are summarized as follows:

> . . . the decline of mortality in the second half of the nineteenth century was due wholly to a reduction of deaths from infectious diseases; there was no evidence of a decline in other causes of death. Examination of the diseases which contributed to the decline suggested that the main influences were: (a) rising standards of living, of which the most significant feature was a better diet; (b) improvements in hygiene; and (c) a favorable trend in the relationship between some microorganisms and the human host. *Therapy made no contributions, and the effect of immunization was restricted to smallpox which accounted for only about one-twentieth of the reduction of the death rate* [McKeown et al., 1975: 391].

While McKeown's interpretation is based on the experience of England and Wales, he also examined the situation in four other European countries; Sweden, France, Ireland, and Hungary (McKeown et al., 1972). His interpretation appears to withstand this examination. As for the twentieth century (1901-71 is the period actually considered), McKeown argued that about three-quarters of the decline were associated with infectious diseases and the remainder with conditions not attributable to microorganisms. He distinguished the infections according to their modes of transmission (via air, water, and food) and isolated three types of influences that figure during the period considered: medical measures (specific therapies and immunization), reduced exposure to infection, and improved nutrition. His conclusion is that

the main influences on the decline in mortality were improved
nutrition on air-borne infections, reduced exposure (from better hygiene) on water- and food-borne diseases and, less
certainly, immunization and therapy on the large number of
conditions included in the miscellaneous group. Since these
three classes were responsible respectively for nearly half,
one-sixth, and one-tenth of the fall in the death rate, it is
probable that the advancement in nutrition was the major
influence [McKeown et al., 1975: 422].

More than twenty years of research by McKeown and his
colleagues culminated in two books — *The Modern Rise of
Population* (1976) and *The Role of Medicine: Dream, Mirage
or Nemesis* (1976) — which draw together his many contributions.

No one in the United States has pursued the thesis with
anything like the same rigor and consistency that characterizes
the work by McKeown and his colleagues in Britain. Around
the early thirties there were several discussions of the questionable effect of medical measures on selected infectious
diseases like diphtheria (Lee, 1931; Wilson and Miles, 1946;
Bolduan, 1930) and pneumonia (Pfizer and Company, 1953). In
a presidential address to the American Association of Im-
munologists in 1954, Magill marshaled an assortment of the
data then available — some from England and Wales — to cast
doubt on the plausibility of existing accounts of the decline in
mortality for several conditions (Magill, 1955). Probably the
most influential work in the United States is that by Dubos,
who, principally in *Mirage of Health* (1959), *Man Adapting*
(1965), and *Man, Medicine, and Environment* (1968) focuses on
the nonmedical reasons for changes in the overall health of
populations. Kass (1971), employing data from England and
Wales, argued that most of the decline in mortality for most
infectious conditions occurred prior to the discovery of either
the "cause" of the disease or some purported "treatment"
for it.

Largely on the basis of clinical experience with infectious
diseases and data from Massachusetts, Weinstein (1974), while
conceding there are (a) some effective treatments that seem to

yield a favorable outcome (e.g., poliomyelitis, tuberculosis, and possible smallpox), argued that (b) despite the presence of supposedly effective treatments some conditions may have increased (e.g., subacute bacterial endocarditis, streptococcal pharyngitis, pneumococcal pneumonia, streptococcal impetigo, gonorrhea, syphilis, herpes progenitalis, and probably one form of bacterial meningitis) and (c) mortality for some conditions shows improvement in the absence of any treatment (e.g., chickenpox).

With the appearance of his book, *Who Shall Live?* Fuchs (1974), a health economist, contributed to a resurgence of interest in the relative contribution of medical care to modern mortality decline in the United States. He believes there has been an unprecedented improvement in health in the United States since about the middle of the eighteenth century that is associated primarily with a rise in real income. While agreeing with much of Fuchs's thesis, we challenge his belief that "beginning in the mid 30's, major therapeutic discoveries made significant contributions independently of the rise in real income" (1974: 54).

AIMS

Our general intention in this chapter and elsewhere is to sustain the ongoing debate on the questionable contribution of medical measures and the presence of medical services to the observable decline in mortality and morbidity in the United States during the twentieth century. Furthermore, from recently available national data we challenge the generally accepted thesis that the health of the nation is slowly but surely improving. More specifically, the following four tasks will be addressed: (a) Correctly calculated age- and sex-adjusted mortality rates (standardized to the population of 1900) for the United States in the twentieth century are presented and discussed in relation to a number of specific and supposedly effective medical interventions (both chemotherapeutic and prophylactic). (b) We find that the health status of the general population (when more appropriately measured by some indicator of morbidity alone, or by a combined measure of both morbidity and mortality) shows no trend toward improvement at all. Indeed, there are reasonable grounds for arguing that the nation's health status may actually be deteriorating. (c) We consider in what way, if any, measures of health outcomes have been altered as a consequence of changes in the utilization of and access to medical care. Finally, (d) some social policy implications are outlined.

THE MODERN DECLINE IN MORTALITY

Although mortality rates for certain conditions, for select age and sex categories, continue to fluctuate or even increase (U.S. Department of Health, Education and Welfare, 1964; Moriyama and Gustavus, 1972; Lilienfeld, 1976), there can be little doubt that a marked decline in overall mortality for the United States has occurred since about 1900 (the earliest point for which reliable national data are available). That such a decline cannot be construed as necessarily reflecting any absolute or relative improvement in the health of the population is discussed below. That it cannot, for the most part, be attributed to the presence of effective medical measures or services (as traditionally conceived and excluding public health) is our immediate concern.

Just how dramatic the modern decline has been in the United States is illustrated in Figure 4.1, which shows age-adjusted mortality rates (standardized to the population of 1900) for males and females separately. Both sexes exhibit a marked decline since 1900. The female decline began to level off by about 1950, while 1960 witnessed the beginning of a slight increase for males. Figure 4.1 also reveals a slight but increasing divergence between male and female mortality since about 1920.

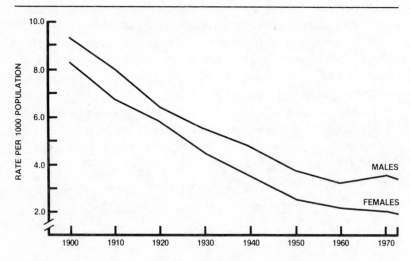

NOTE: For these and all other age- and sex-adjusted rates in this chapter, the standard population is that of 1900.

Figure 4.1   Trend in Mortality for Males and Females Separately (using age-adjusted rates) for the United States, 1900-73

Figure 4.2 depicts the decline in the overall age- and sex-adjusted mortality rate standardized to the population structure of 1900 (SDR) that has been in evidence in the United States since the beginning of the century. Between 1900 and 1973, there was a 69.2 percent decrease in overall mortality. The average annual rate of decline from 1900 until 1950 was .22 per 1,000, after which it became an almost negligible decline of .04 per 1,000 annually. Of the total fall in the standardized death rate between 1900 and 1973, 92.3 percent occurred prior to 1950. Figure 4.2 also plots the decline in the standardized death rate *after* the total number of deaths in each age and sex category has been reduced by the number for deaths attributed to eleven major infectious conditions (namely, typhoid, smallpox, scarlet fever, measles, whooping cough, diphtheria, influenza, tuberculosis, pneumonia, diseases of the digestive system, and poliomyelitis). It should be noted that, although this latter rate also shows a decline (at least until 1960), its slope is much more shallow than that for the overall standardized death rate, indi-

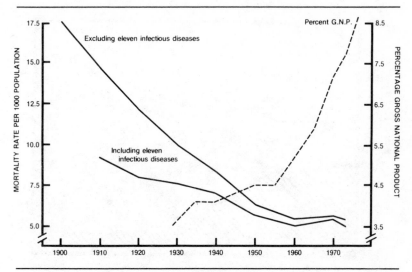

Figure 4.2   Age- and Sex-Adjusted Mortality Rates for the United States, 1900-73, Including and Excluding Eleven Major Infectious Diseases, Contrasted with the Proportion of GNP Expended on Medical Care

cating that a major part of the decline since about 1900 may be attributed to the disappearance of infectious type conditions. An absurdity is reflected in the third broken line in Figure 4.2, which also plots the increase in the proportion of the gross national product expended annually for medical care. *It is evident that the beginning of the precipitate and still unrestrained rise in medical care expenditures began when nearly all (92 percent) of the modern decline in mortality this century had already occurred.*

Figure 4.3 illustrates how the proportion of deaths contributed by infectious and chronic conditions has changed in the United States since the beginning of the twentieth century. In 1900, about 40 percent of all deaths were contributed by eleven major infectious diseases, 16 percent by three main chronic conditions, 4 percent by accidents, and the remainder (37 percent) by other causes. By 1973, only 6 percent of all deaths were due to these eleven infectious diseases, 58 percent to the same three chronic conditions, 9 percent to accidents, and 27 percent to other causes.[1]

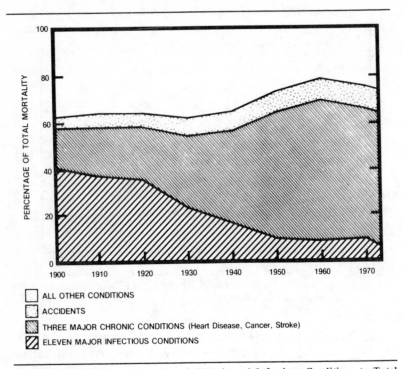

Figure 4.3   Changing Contribution of Chronic and Infectious Conditions to Total Mortality (age- and sex-adjusted) in the United States, 1900-73

Now to what phenomenon or combination of events can we attribute this modern decline in overall mortality? Who (if anyone), or what group, can claim to have been instrumental in effecting this reduction? Can anything be gleaned from an analysis of experience with mortality to data that will inform social policy for the future?

It should be reiterated that a major concern here is to determine the effect, if any, of the presence of medical measures (both chemotherapeutic and prophylactic) on the decline of mortality in the United States. It is clear from Figures 4.2 and 4.3 that most of this observable decline is due to the rapid disappearance of some of the major infectious diseases. Since this is where most of the decline has occurred, it is logical to focus a study of the effect of medical measures on this category of conditions. Given an overall increase in mortality since about 1900 for the major chronic conditions, it is illogical to investigate the possible contribution of medical measures to reducing mortality for these conditions, when there has been no clear improvement.

THE EFFECT OF MEDICAL MEASURES
ON TEN DISEASES THAT HAVE DECLINED

Table 4.1 summarizes data on the effect of major medical interventions (both chemotherapeutic and prophylactic) on the decline in the age- and sex-adjusted death rate (standardized to the population of 1900) for ten major infectious diseases in the United States between 1900 and 1973. Together, these diseases account for approximately 30 percent of all mortality at the turn of the century and nearly 40 percent of the total decline since then. The ten diseases were selected on the following criteria: (a) that some decline in the death rate had occurred in the period 1900-1973; (b) that a significant decline in the death rate is commonly attributed to some specific medical measure for the disease; and (c) that adequate data for the disease over the period 1900-1973 were available. The diseases of the digestive system were omitted primarily because of lack of clarity in diagnosis of specific diseases such as gastritis and enteritis.

Some additional points of explanation should be noted in relation to Table 4.1. First, the year of medical intervention coincides (as nearly as can be determined) with the first year of widespread or commercial use of the appropriate drug or vaccine. This date does *not* necessarily coincide with the date the measure was either first discovered or subject to clinical trial. Second, the decline in the death rate for smallpox was calculated using the death rate for 1902 as being the earliest year for which this statistic is readily available (U.S. Bureau of the Census, 1906). For the same reasons, the decline in the death rate from poliomyelitis was calculated from 1910. Third, the

TABLE 4.1 Contribution of Medical Measures (both chemotherapeutic and prophylactic) to the Fall in the Age and Sex-Adjusted Death Rates (SDR) of Ten Common Infectious Diseases, and to the Overall Decline in the SDR for the United States, 1900-73

Disease	Fall in SDR per 1,000 Population, 1900-1973 (a)	Fall in SDR as % of the Total Fall in SDR $(b) = \frac{(a)}{12.14} \times 100\%$	Year of Medical Intervention (either chemotherapy or prophylaxis)	Fall in SDR per 1,000 Population After Year of Intervention (c)	Fall in SDR After Intervention as % of Total Fall for the Disease $(d) = \frac{(c)}{(a)} \times 100\%$	Fall in SDR After Intervention as % of Total Fall in SDR for All Causes $(c) = \frac{(b)(c)}{(a)}\%$
Tuberculosis	2.00	16.48	Izoniazid/Streptomycin, 1950	0.17	8.36	1.38
Scarlet fever	0.10	0.84	Penicillin, 1946	0.00	1.75	0.01
Influenza	0.22	1.78	Vaccine, 1943	0.05	25.33	0.45
Pneumonia	1.42	11.74	Sulphonamide, 1935	0.24	17.19	2.02
Diphtheria	0.43	3.57	Toxoid, 1930	0.06	13.49	0.48
Whooping cough	0.12	1.00	Vaccine, 1930	0.06	51.00	0.51
Measles	0.12	1.04	Vaccine, 1963	0.00	1.38	0.01
Smallpox	0.02	0.16	Vaccine, 1800	0.02	100.00	0.16
Typhoid	0.36	2.95	Chloramphenicol, 1948	0.00	0.29	0.01
Poliomyelitis	0.03	0.23	Vaccine, Salk/Sabin, 1955	0.01	25.87	0.06

table shows the contribution of the decline in each disease to the total decline in mortality over the period 1900-1973 (column b). The overall decline during this period was 12.14 per 1,000 population (17.54 in 1900 to 45.39 in 1973). Fourth, in order to place the experience for each disease in some perspective, Table 4.1 also shows the contribution of the relative fall in mortality after the intervention to the overall fall in mortality since 1900 (column e). In other words, the figures in this last column represent the percentage of the total fall in mortality contributed by each disease after the date of medical intervention.

It is clear from column b that only tuberculosis and pneumonia contributed substantially to the decline in total mortality between 1900 and 1973 — 16.5 percent and 11.7 percent respectively. The remaining eight conditions *together* contributed less than 12 percent to the total decline over this period. Disregarding smallpox (for which the only effective measure was introduced about 1800), only influenza, whooping cough, and poliomyelitis show what could be considered substantial declines (25 percent or more) after the date of medical intervention. However, it should be noted that, even under the somewhat unrealistic assumption of a constant (linear) rate of decline in the mortality rates, only whooping cough and poliomyelitis even approach the percentage that would have been expected anyway. The remaining six conditions (tuberculosis, scarlet fever, pneumonia, diphtheria, measles, and typhoid) showed negligible declines in their mortality rates subsequent to the date of medical intervention. The seemingly quite large percentages for pneumonia and diphtheria (17.2 and 13.5, respectively) must of course be viewed in the context of relatively early interventions — 1935 and 1930.

The detailed mortality trends for these diseases and the relation of these trends to the medical interventions have been thoroughly examined elsewhere (McKinlay and McKinlay, 1977). It was found that for tuberculosis, typhoid, measles, and scarlet fever, the medical measures considered were introduced at the point when the death rate for each of these diseases was already negligible. Any change in the rates of decline that may have occurred subsequent to the interventions could be only minute. Of the remaining five diseases considered (excluding smallpox with its negligible contribution), it was only for poliomyelitis that the medical measure appears to have produced any noticeable change in the trends. Given peaks in the death rate for 1930, 1950 (and possibly for 1910), a comparable peak could have been expected in 1970. Instead, the death rate dropped to the point of disappearance after 1950 and has remained negligible. The four other diseases (pneumonia, influenza, whooping cough, and diphtheria) exhibit relatively smooth mortality trends that are unaffected by the medical measures, even though these were introduced relatively early, when the death rates were still notable.

From the evidence considered, only poliomyelitis appeared to have a noticeably changed death rate subsequent to intervention. Assuming (unrealistically) that this change was entirely due to the vaccines, then only about 1 percent of the decline following interventions for these diseases considered here (column d of Table 4.1) could be attributed to medical measures. Rather more conservatively, if we attribute some of the subsequent fall in the death rates for pneumonia, influenza, whooping cough, and diphtheria to medical measures, then, perhaps 3.5 percent of the fall in the overall death rate can be explained through medical intervention in the major infectious diseases considered here. Indeed, given that it is precisely for these diseases that medicine claims most success in lowering mortality, 3.5 percent probably represents a reasonable upper-limit estimate of the total contribution of medical measures to the decline in mortality in the United States since 1900.

HOW RELIABLE ARE MORTALITY STATISTICS?

It would be foolhardy indeed to dismiss all studies based on mortality measures simply because they are possibly beset with *known limitations*. One can argue that such data are preferable to those the limitations of which are either unknown or, if known, unable to be estimated. Because of an overawareness of potential inaccuracies, there is a tendency to disregard or devalue studies based on mortality evidence — even though there are innumerable examples of their fruitful use as a basis for planning and informed social action (Alderson, 1976). Austin Bradford Hill (1955) considers one of the most important features of Snow's work on cholera to be his adept use of mortality statistics. Another example is the study by Inman and Adelstein (1969) of the circumstantial link between the excessive absorption of bronchodilators from pressurized aerosols and the epidemic rise in asthma mortality in children aged 10 to 14 years. Moreover, there is evidence that some of the inaccuracies of mortality data tend to cancel each other out.[2] Consequently, while mortality statistics may be unreliable for use in individual cases, when pooled for a country and employed in population studies, they have revealed important trends (as in the above discussion) generated fruitful hypotheses and already resulted in informed social action.

Schneyer et al. (1980) have suggested that this analysis is misleading and results in an erroneous devaluation of the importance of medical measures in the decline of mortality in this century. Their argument is based on a questionable broadening of the definition of "medical measures" to include public health measures such as the sanitation of the water supply. Public health measures are seen as a product of biomedical research and thus as in the same category as specific chemotherapeutic and prophylactic therapies. They fail to explain declines in disease-specific mortality that preceded knowledge of the etiology and/or cure of the disease or to acknowledge the major contributions of improved nutritional status (McKeown et al., 1975: 422) or increases in the real income of the population (Fuchs, 1974). Furthermore, the failure to distinguish between public health measures and specific medical interventions leads to a conclusion that provides little clear direction for policymakers. Our international fiscal crisis (McKinlay, 1977; O'Connor, 1973) requires criteria with which to inform government upon priorities for the allocation of scarce resources.

Therefore, the relative contributions of public health measures, medical interventions, and biomedical research need to be assessed.

It should be noted that whatever limitation and risks may be associated with mortality statistics, they obviously apply equally to all studies that employ them — both those attributing the decline in mortality to medical measures and those arguing the converse or something else entirely. If such data constitute acceptable evidence in support of the presence of medicine (which indeed they are), then it is not unreasonable or illogical to employ them in support of some opposing position. One difficulty is that, depending on the nature of the results, double standards of rigor seem to operate in the evaluation of different studies. Not surprisingly, those that challenge prevailing myths are subject to the most stringent methodological and statistical scrutiny, while supportive studies, which frequently employ the flimsiest impressionistic data and inappropriate techniques of analysis, receive general and uncritical acceptance. Even if all possible "ideal" data were available (which they never will be) and if, after appropriate analysis, they happened to support the viewpoint of this chapter, we are doubtful that medicine's protagonists would find our thesis any more acceptable.

Since mortality measures have well-known limitations and obvious risks are entailed in their extensive use, some cautionary caveats concerning the data and mode of analysis of this chapter may be appropriately introduced here. We fully appreciate (a) difficulties introduced by changes in the registration area in the United States in the early twentieth century; (b) that often no single disease, but a complex of conditions, may be responsible for death (Krugger, 1966); (c) that studies reveal considerable inaccuracies in statements of the cause of death (Moriyama et al., 1958); (d) that there are changes in what it is fashionable to diagnose (for example, ischemic heart disease and cerebrovascular disease); (e) that changes in disease classifications (Dunn and Shackley, 1945) make it difficult to compare some conditions over time and between countries (Reid and Rose, 1964); (f) that some conditions result in immediate death while others have an extended period of latency; and (g) that many conditions are severely debilitating and consume vast medical resources but are now generally nonfatal (e.g., arthritis and diabetes). Other obvious limitations could be added to this list.

Table 4.2 illustrates the inability of mortality rates to reflect accurately current disease in the U.S. population. Although disease classifications are not entirely consistent, it is clear that the five major causes of death are not the five major causes of activity limitation, physician contact or hospital admission. The categories "bodily impairment" and "acute injuries" include, in large part, the acute and residual phases of nonfatal accidents. Thus, if we can include these under the general heading of "accidents," this and heart disease are the only categories that consistently appear among the top five diagnoses. (Heart disease is included under physician visits as a subset of "chronic circulatory conditions.") Chronic respiratory diseases are preempted for women by complications of childbearing as a reason for hospital admission. As a further indication of the divergence between the major causes of death and the major causes of ill health and disability in the general population, one can consider the percentage of all diagnoses, for any of the four events, contributed by the five major causes of death. While these five diagnoses account for 76 percent of all mortality, they account for less than 40 percent of short-stay hospital discharges (36.3 percent for males, 22.1 percent for females), about 20 percent of physician visits, and, as far as can be estimated, about 20 percent of all activity limitation. Although this last percentage is an underestimate (due to the exclusion of accidents as a cause) the correct percentage is probably of the same order as that for the other two events.

CHANGES IN HEALTH STATUS

Clearly, trends in mortality statistics are not sufficient to describe the changing health status of the general population,

TABLE 4.2 Contribution of Five Primary Diagnoses to Mortality and Three Related Events: United States, 1974-78[a]

Mortality 1978	% Total	Limitation of Activity[b] 1974 Male	% Total	Female	% Total	Physician Visits 1975	% Total	Short-Stay Hospital Discharges 1975 Male	% Total	Female	% Total
Diseases of the heart	37.8	diseases of the heart	18.0	arthritis and rheumatism	19.6	acute respiratory conditions	15.1	digestive system diseases	4.9	complications of childbearing and puerperium	19.6
Malignant neoplasms	20.6	bodily impairment (except paralysis)	16.6	diseases of the circulatory system (except heart)	17.6	acute injuries	11.9	accidents	14.6	genitourinary diseases	11.9
Cerebrovascular disease	9.0	chronic respiratory disease	15.0	diseases of the heart	14.5	chronic circulatory conditions	10.0	diseases of the heart	11.1	digestive system diseases	10.9
Accidents	5.6	diseases of the circulatory system (except heart)	11.4	bodily impairment (except paralysis)	14.5	chronic respiratory conditions	7.1	genitourinary diseases	7.6	accidents	7.8
Influenza/pneumonia	3.0	arthritis and rheumatism	10.1	chronic respiratory disease	14.5	acute infective and parasitic conditions	4.7	malignant neoplasms	5.3	diseases of the heart	6.5
Total contribution of 5 leading causes of mortality to each event	76.0		23.0[c]		19.2[c]		20.3		36.3		22.1

a. Sources: In all cases the most recent available data have been used (National Center for Health Statistics, *Monthly Vital Statistics Report*, Vol. 27, No. 13, August 1979, Hyattsville, MD, DHEW; National Center for Health Statistics, Health Interview Survey, 1974, unpublished data; National Center for Health Statistics, *Vital and Health Statistics*, Series 13, No. 35 and Series 10, No. 128, Hyattsville, MD, DHEW; National Center for Health Statistics, *Nursing Home Survey*, 1974, unpublished data).

b. Condition groups are not mutually exclusive and the proportions are, therefore, slightly overestimated.

c. These percentages do not include accidents or influenza/pneumonia, as those are not listed under conditions causing chronic limitations.

and some additional measure of morbidity is desirable. One obvious approach is to measure the prevalence of certain chronic conditions and/or the incidence of some common infectious diseases and observe any changes in these rates over time. This method, however, is beset with problems of changing disease patterns (a disease that was common in 1930 may have virtually disappeared by 1970), changing fashions in diagnosis (for example, the appearance of "chronic obstructive lung disease" as a diagnosis in recent years), and the difficulties of measuring the prevalence of some conditions with long, asymptomatic latency periods (heart disease, cirrhosis of the liver, some kidney diseases, many cancers, and so on). Another approach is to consider indirect measures of illness, such as days of restricted activity, workday loss, or the prevalence of some carefully defined type of disability. Unfortunately, in most countries, mortality data are all that are available for a sufficient time span. In the United States, however, data on morbidity have been collected routinely in continuing national surveys for two decades (1957 onward). These unique data provide the indirect measure of illness that can be used in conjunction with mortality statistics to provide a more sensitive measure of trends in the nation's health.

AVAILABLE MORBIDITY DATA

When one is employing indirect measures such as work time lost or activity restriction, which can be uniformly defined and which are not dependent on changing disease patterns for comparability, many of the major problems associated with the use of mortality data are overcome. Not surprisingly however, other difficulties are introduced. For example, loss of workdays may be predominantly a function of benefit policies, changes in the type of work, morale on the job, and seasonal variations rather than the presence of some medical condition, and thus may not measure morbidity at all (Alderson, 1967; Morris, 1965; Taylor, 1941).

Available data are not entirely free from the types of problems mentioned, and an effort is made here to compensate wherever possible for such biases. Also, because morbidity from other than the reportable infectious diseases has been measured regularly only since the National Center for Health Statistics' Health Interview Survey began in the United States in 1957, only between one and two decades of data are available from which to determine time trends (depending on the type of data required).[3] The survey estimates are further restricted to the civilian, noninstitutionalized population, necessitating the use of supplemental data (where available) for the institutionalized population.

Several measures of aspects of disability have been employed in these surveys, including bed disability days and restricted activity days (for both acute and chronic conditions), limitation in activity, and limitation in mobility (for chronic conditions). In order to include measures of disability for both acute and chronic conditions using comparable definitions, restricted activity days and limitation in activity were chosen, as both refer to curtailment of usual activity to some degree.[4] Although these definitions were approximately comparable, the estimates generated are different and not mutually exclusive, reflecting the distinct nature of acute episodes and chronic conditions.

One obvious way of combining these two measures is to consider the number of restricted activity days per person, per year, separately for those with no activity limitation and for those limited in or unable to perform an activity because of some chronic condition. A combined estimate of restricted activity days was formed, incorporating data from nursing home populations, in order to provide some estimate (however approximate) for the total civilian population. This was done assuming that those confined to nursing home care could reasonably be considered as restricted in activity 365 days per year. These combined estimates show a consistent increase in restricted activity days per person, per year, of less than one day (for those under 45 years) and as much as five days (for those 45 years or older).

It is possible that the higher rates of limitation in men are at least partly due to the rise in benefits, but given the marked increases in limitation among women and children (those under 17 years), it is reasonable to conclude that there has been a comparable real increase in disability in men. As an illustration, for women not working in the age group 17-44 years, there was a 48 percent increase in limitation over the 15 years considered, compared to a 56 percent increase among men in the same category. Thus, if we can assume a comparable rate of increase in disability, then only about one-sixth of the increase in this group of men would be attributable to the effect of benefits.

One major problem of the effect of benefits on disability rates has been investigated as far as data permit, and it appears that the effect can probably explain only a small proportion of the increase. A second problem in comparing rates over time is the effect of changes and improvements in data collection methodology, since the National Center for Health Statistics has revised its survey techniques over the years. It can be argued that the apparent increase in disability rates is largely a function of improvements in survey methodology in that more respondents may be more correctly reporting limitation in activity now than they did in earlier years, while the true rates have remained essentially the same. This objection can be neither dismissed nor confirmed easily, as any improvements in reporting are very difficult to measure. However, by restricting our analysis to limitation in *major activity only* as a measure of disability, we have focused on an extreme group who are more likely to be clearly defined — despite occasional changes in the way such information is solicited. It can certainly be argued that greater ambiguity is likely to exist concerning limitation in other than major activity. A third source of concern relates to the reliability of self-reported disability data. However, it is now well documented that people tend to underreport illness and associated problems to a marked extent (Cannell et al., 1977). Thus, even given the possible effects of benefits and methodological improvements, we are almost certainly still considering underestimates of the real level of disability in the general population.

THE PROBABILITY OF A LIFE
FREE OF DISABILITY

If we conclude that there has been a real increase in disability, at least between 1959 and 1975, the next logical step is to combine this information meaningfully with mortality data to provide some more sensitive summary indicator of the health of the general population. Sullivan (1971) has proposed a simple and useful index of mortality and morbidity that consists of the calculation of *a life expectancy free of disability*. This is accomplished simply by multiplying the probability of survival by the estimated conditional probability of remaining free of disability (assuming survival). The resulting probability of survival free of disability is then used (instead of simply the probability of survival) to calculate an expected life span in the usual way. The resulting average is of life free of disability, which can then be compared with the overall life expectancy.

The conditional probabilities of remaining free of disability were calculated as follows: Because long-term or permanent disability is more closely measured by limitation in activity due to chronic conditions than by restricted activity days, the proportion of the population either limited in or unable to carry out a major activity, or residing in a nursing home, was estimated for 1964 and 1974 using the data sources cited above. Since this included only the more severely limited and because this limitation was due to a chronic condition, it seemed reasonable to assume that this limitation persisted throughout the year. The proportion so calculated, therefore, could be considered as an estimate of the probability that a person alive at the beginning of the year would be disabled during the year.

The resulting life expectancies for 1964 and 1974 are presented in Table 4.3. From the far right-hand columns of this table, it is clear that, *although overall life expectancy has increased over this decade, almost all of this increase was in years of disability.* For example, in 1964, a man reaching the age of 45 could expect to live nearly 20 more years free of disability, and a further 7.2 years disabled. In 1974, a man of the same age

TABLE 4.3 **Trend in Total Life Expectancy and Life Expectancy Free of Disability for Various Ages, Male and Female, in the United States, 1964-74**

Sex	Age (years)	1964			1974			Change	
		Total L.E.	L.E. Free of Disability	L.E. with Disability	Total L.E.	L.E. Free of Disability	L.E. with Disability	Increase in Total L.E.	Increase in L.E. Free of Disability
Male	0	66.8	59.2	7.6	68.1	59.2	8.9	1.3	0.0
	45	27.1	19.9	7.2	27.8	19.8	8.0	0.7	-0.1
	65	12.8	6.6	6.2	13.4	7.2	6.2	0.6	0.6
Female	0	73.7	65.5	8.2	75.8	65.3	10.5	2.1	-0.2
	45	32.5	25.0	7.5	33.9	24.6	9.3	1.4	-0.4
	65	16.2	10.2	6.0	17.5	10.7	6.8	1.3	0.5

SOURCES: U.S. Bureau of the Census, *Mortality Statistics, Part II* (Washington, DC: U.S. Government Printing Office, 1964); U.S. Bureau of the Census, *Vital Statistics, 1974*; National Center for Health Statistics, *Health Interview Survey, 1964*; C.S. Wilder (1976), Table 1; National Center for Health Statistics, *Nursing Home Survey, 1974*; Nelson (1967).

had about the same number of years to live free of disability, but 8 years to live disabled. In other words, the gain in overall life expectancy over the decade of 0.7 years, for men aged 45 years, was a gain in years of disability only. The same argument applies for all age and sex categories — except for those over 65 years, who appear to have experienced small gains in expected life free of disability.

Given that the estimates of life expectancy free of disability are conservative (excluding data on those limited in other than major activity) then the recent gains in life expectancy may have been merely gains in years of disability, and there may even have been some decrease in years free of disability.

THE INVERSE CARE LAW

In 1972, Julian Tudor Hart proposed the Inverse Care Law ("that the availability of good medical care tends to vary inversely with the need for it in the general population served") to describe the British National Health Service's failure to achieve its goal of reducing inequalities. Contrary to the work of Rein (Rein, 1969a, 1969b), Hart observed that many of the indicators of mortality and morbidity showed a clear class gradient, and he pointed to the work of Titmuss (1968) for additional corroborative evidence.

Hart's observation, like those of several other health policy analysts, suggests that health is not randomly distributed throughout society. In general, it is agreed that the poor, certain minority groups (black, in particular), and the elderly are more susceptible to particular forms of morbidity and mortality and, furthermore, that there are patterns of differential treatment in existence as a function of, among other things, age, race, social class, and area of residence. Syme and Berkman have found that low socioeconomic status is associated with higher rates of infectious and parasitic diseases, lower life expectancy, and higher mortality rates from all causes — a finding first observed in the twelfth century (Syme and Berkman, 1981). Minority group membership has also been linked with consistently higher rates of general and infant mortality, lower expectation of life at birth and beyond, and higher morbidity rates (disability days and chronic conditions) than is the case for whites (Syme and Berkman, 1981; U.S. Department of Health, Education and Welfare, 1979; Weeks, 1977). When poverty and minority group status are added to the fact that one is old and/or lives in either an urban ghetto or in a rural area, individual sociodemographic characteristics may be said to coalesce into "multiple jeopardy."

Though we have witnessed rising living standards over the last century, substantial differences in health status remain. It is conceivable that the problems of the poor, minority groups, and the aged are the result of a complex of reduced access and/or failure to benefit from medical care, coupled with living in a toxic, hazardous environment that is pathogenic socially, medically, and psychologically insofar as it increases vulnerability to specific diseases and morbid conditions.

The second part of Hart's thesis is that the availability of health services (both quantity and quality of care) is a function of nonmedical criteria. In other words, here he speaks of the notion of equality of access to care. Though many people have written rather extensively on this topic (see, for example, Aday and Anderson, 1981; Anderson et al., 1975, 1976; Goodrich et al., 1970; Wyszewianski and Donabedian, 1981) and others have written, attesting to the virtues of equality/equity in presumably egalitarian societies such as the United States (Titmuss, 1968; Crichton, 1980; Syme and Berkman, 1981), a number of studies reveal that even the introduction of government-sponsored health insurance is no guarantee that there will be an equitable distribution of health care resources, much less a guarantee that it will produce equality of health outcomes. As Crichton has remarked concerning the Canadian health care system, equality "is a value easier to legislate than to accomplish" (Crichton, 1980: 248).

We conclude that the inverse care law continues to apply in the United States despite more than a decade of increased government intervention. Although it would be incorrect to generalize from any one report, a persistent inverse relationship between socioeconomic and/or racial minority status and various indicators of health has been reported throughout the 1970s in nationwide data from the National Center for Health Statistics, as well as in other, smaller scale studies (such as Goodrich et al., 1970; Diehr et al., 1979).

A wide range of factors are often included in analyses of access to medical care (Aday and Anderson, 1981). These have included *structural* indicators, such as the volume and distribution of hospital and clinic facilities and health manpower; *process* indicators, which tap the ease in obtaining care; and *subjective* (satisfaction) and *objective* (utilization) indicators. According to many of these studies, considerable gains, particularly as measured by utilization data, have been made by the poor since the introduction of Medicaid in the mid-1960s (Davis, 1976; Davis et al., 1981; Donabedian, 1976; Somers and Somers, 1977; Wilson and White, 1977). Table 4.4 illustrates current rates of utilization of physician and hospital services by blacks and those in the lowest income categories.

The inverse care law, however, describes inequality of access to *good* medical care, raising issues of quality as well as quantity of care. Townsend's (1974) comments regarding British health and human services seem particularly cogent:

> In building up a picture of utilization of different health services it must not be supposed, because some services are heavily utilized by the poor working classes, that this is necessarily contributory evidence of equitable provision of health services as a whole. Like other institutional systems of society, the health system is organized in a hierarchy of values and status. *No one today would argue that the heavy utilization of secondary-modern schools by the working classes constitutes evidence of equality of educational provision* [emphasis added].

Table 4.5 shows the location of physician visits by race and income levels. Others have documented the problems of medical care received in outpatient departments and emergency rooms: discontinuous care, substandard facilities, complicated bureaucratic procedures, and difficulties in communication be-

TABLE 4.4   Utilization of Physician and Hospital Services
by Income and Race, 1979

	Physician Visits[a]	Hospital Discharges[b]
	(per 1,000 population)	
Family Income[c]		
$7,000	5,358.2	163.0
$7,000-$9,999	5,139.7	139.2
$10,000-$14,999	4,582.1	127.4
$15,000-$24,999	4,624.6	110.1
$25,000+	4,658.3	107.2
Race[c]		
Black	4,764.3	137.3
White	4,528.0	120.9
Ratio		
Income (lowest/highest)	1.15	1.52
Race (black/white)	1.05	1.14

SOURCE: U.S. Department of Health and Human Services, *Health: United States, 1981*.
a. All sources/places of care.
b. Nonfederal short-stay hospitals.
c. Age adjusted by the direct method to the 1970 civilian noninstitutionalized population using four age intervals.

tween doctor and patient (Duff, 1976; Sidel and Sidel, 1977; Stoeckle, 1976). According to these data, a greater percentage of blacks and individuals with low income receive care in outpatient departments and emergency rooms.

Tables 4.6 and 4.7 provide further evidence of this disparity in the availability of quality service. Table 4.6 demonstrates that members of racial minorities and low-income groups are more likely to travel longer to a regular source of care, and Table 4.7 shows that they will wait longer to be seen when they get there. These are major barriers, not mere inconveniences, as Medicaid does not always provide reimbursement for or assistance in obtaining transportation, child care, or time off from work.

In addition, important disparities in the types of services utilized have persisted. While considerable improvement in equalizing utilization rates of physician and hospital services

TABLE 4.5   Location of Physician Visits by Income and Race, 1979

| | Source or Place of Care | | |
	Doctor's Office, or Group Group Practice	Hospital Outpatient Department or Emergency Room	Telephone
	(visits per 1,000 population)		
Family Income[a]			
$7,000	3141.6	1128.4	543.9
$7,000-$9,999	3445.3	790.9	570.5
$10,000-$14,999	3035.9	618.1	631.5
$15,000-$24,999	3253.5	519.3	637.4
$25,000+	3349.9	425.4	624.1
Race[a]			
Black	2641.3	1110.5	348.5
White	3278.9	556.4	650.1
Ratio			
Income (lowest/highest)	.94	2.65	.87
Race (black/white)	.80	2.00	.54

SOURCE: U.S. Department of Health and Human Services, *Health: United States, 1981.*

a. Age adjusted by the direct method to the 1970 civilian noninstitutionalized population using four age intervals.

has been obtained, lower utilization of dental and preventive services by members of low-income and racial minority groups remain.

Table 4.8 shows utilization rates for dental services. Blacks and low-income groups get less dental care than do whites and those in the highest income groups — in sharp contrast to the greater need among these groups found by Luft (1978) in his summary of the results of several National Health Examination Surveys. Similar findings exist in comparisons of utilization of preventive health services. For example, a greater percentage of whites and middle-income individuals report a general checkup in the preceding year. Also, immunization levels for children for rubella, polio, and DPT are lower for nonwhite children (U.S. Department of Health, Education and Welfare, 1979).

**TABLE 4.6   Travel Time to Regular Source of Care
by Income and Race**

	% 15 min.	% 15-29 min.	% 30-59 min.	% 60+ min.
Income[a]				
Below near poverty / Above near poverty	.78	1.03	2.28	4.00
Race[b]				
Racial minority / White	.76	1.09	1.35	1.54

a. Aday and Anderson (1974).
b. National Center for Health Statistics, 1980.

**TABLE 4.7   Waiting Time at Regular Source of Care
by Income and Race, 1974**

	% Immediate	% 1-30 min.	% 31-60 min.	% 60+ min.
Income				
Below near poverty / Above near poverty	.42	.70	1.17	1.83
Race				
Racial minority / White	.50	.68	1.12	2.06

SOURCE: Aday and Anderson (1974).

Most important, government interventions (e.g., Medicare and Medicaid) have failed to equalize health outcomes. On several measures of health status, members of racial minorities and lower income groups continue to show greater need.

While age-adjusted mortality rates have dropped for the whole population since 1940, nonwhite males and females have higher rates than their white counterparts. Since 1955, the ratio of nonwhite to white mortality rates for both sexes has remained around 1.40 (U.S. Department of Health, Education and Welfare, 1979). Similar higher mortality rates for blacks as compared to whites are found in specific disease categories. In

TABLE 4.8   Utilization of Dental Services by Income and Race, 1979

	Dental Visits per 1,000 Population
Family Income[a]	
$7,000	1260.9
$7,000-$9,999	1235.3
$10,000-$14,999	1323.0
$15,000-$24,999	2356.6
$25,000+	2356.6
Race[a]	
Black	1036.6
White	1797.7
Ratio	
Income (lowest/highest)	.54
Race (black/white)	.58

SOURCE: U.S. Department of Health and Human Services, *Health: United States, 1981*.
a. Age adjusted by the direct method to the 1970 civilian noninstitutionalized population, using four age intervals.

1977, all of the following conditions showed higher mortality rates for blacks than whites: cardiovascular disease, malignant neoplasms, diabetes, accidents, and homicides (U.S. Department of Health, Education and Welfare, 1980).

Additional measures (Tables 4.9 and 4.10) document this continued inequality in health status. Blacks and members of low-income families report more restricted activity and bed disability days, and a greater percentage of these two groups report chronic activity limitation and self-ratings of health status as poor or fair, as compared to whites and those with higher incomes.

In spite of over a decade of government intervention and the equalization of physician and hospital services utilization, persistent gaps remain in access to *quality* care and health status (see Table 4.11).

What is the reason for this failure in health policy? We offer four explanations. First, inequality of access, based in part on

TABLE 4.9   Restricted Activity Days and Bed Disability Days, 1979

	Restricted Activity Days	Bed Disability Days
	(number per person per year)	
Family Income[a]		
$7,000	30.6	10.6
$7,000-$9,999	22.0	8.3
$10,000-$14,999	17.5	6.4
$15,000-$24,999	15.2	5.5
$25,000+	14.2	5.1
Race[a]		
Black	24.8	9.7
White	18.0	6.3
Ratio		
Income (lowest/highest)	2.15	2.08
Race (black/white)	1.38	1.54

SOURCE: U.S. Department of Health and Human Services, *Health: United States, 1981*.

a. Age adjusted by the direct method to the 1970 civilian noninstitutionalized population, using four age intervals.

structured inequalities in the society, will not respond merely to the removal of financial barriers to care. This conclusion is far from novel and was summarized by Titmuss (1968: 196-197) as follows:

> . . . equality of access is not the same thing as equality of outcome. . . . We have learned from fifteen years' experience of the health service that the higher income groups know how to make better use of the service; they tend to receive more specialist attention, occupy more beds in better equipped and staffed hospitals, receive more elective surgery, have better maternity care, and are more likely to get psychiatric help and psychotherapy than low-income groups — particularly the unskilled. . . . Universalism in social welfare, though a needed prerequisite toward reducing and removing formal barriers of social and economic discrimination, does not by itself solve the problem of how to reach the more-difficult-to-reach with better medical care.

TABLE 4.10   Self-Assessment of Health and Limitation of Activity
by Income and Race, 1979

	Self-Assessment of Health as Fair or Poor	Reporting Limitation of Activity
	(percentage of population)	
Family Income[a]		
$7,000	22.5	22.0
$7,000-$9,999	16.9	17.8
$10,000-$14,999	12.5	14.0
$15,000-$24,999	9.5	11.7
$25,000+	6.7	10.0
Race[a]		
Black	20.3	17.7
White	11.4	13.6
Ratio		
Income (lowest/highest)	3.36	2.20
Race (black/white)	1.78	1.30

SOURCE: U.S. Department of Health and Human Services, *Health: United States, 1981*.
a. Age adjusted by the direct method to the 1970 civilian noninstitutionalized population, using four age intervals.

Second, the welfare/insurance scheme employed by Medicaid (and Medicare) allows too many people to fall through the gaps in coverage and leads to a two-class system of health care (Donabedian, 1976). Together these factors limit access to minimally adequate health care. One has only to look at the confusing array of state regulations for Medicaid to appreciate this. Furthermore, according to Schroeder (1981), about one-fifth of our population has insufficient insurance to cover the costs of a long hospital stay.

Third, the focus on equality of access (quantity of medical care available) has been overemphasized to the exclusion of the more important consideration of the quality of medical care available. Policy based only on increasing utilization will not necessarily improve the health of the population. McKinlay and McKinlay (1977) have demonstrated the minor contribu-

TABLE 4.11   Comparisons of Selected Measures of Access and
Health Status by Income and Race: 1972, 1979

| | Income (lowest/highest) | | Race (black/white) | |
	1972[a]	1979[b]	1972	1979
Physician visits	.97	1.15	.90	1.05
Hospital discharges	1.45	1.52	.95	1.14
Office visits	.90*	.94	.78	.80
OP/ER visits	1.86*	2.65	2.27	2.00
Telephone consultations	.56*	.87	.40	.54
Dental visits	.34	.54	.44	.58
Restricted activity days	2.17	2.15	1.31	1.38
Bed disability days	2.23	2.08	1.42	1.54
Self-assessment fair/poor	4.74	3.36	1.88	1.78
Limitation of activity	2.24	2.20	1.28	1.30

a. Source: U.S. Department of Health, Education and Welfare, *Health: United States, 1979.*
b. Source: U.S. Department of Health and Human Services, *Health: United States, 1981.*
*Figure calculated with data that did not meet standards of reliability and precision of National Center for Health Statistics.

tions made by medical interventions to the decline in mortality in the United States. More specifically, during the 1960s and 1970s, increased utilization of nonobstetrical services in this country did not lead to improvements in health status for adults, according to Benham and Benham (1976). Underlying this policy is the erroneous assumption that we are not making progress against illness because there has not been enough medical manpower, biotechnology, hospital construction, federal investment, and so forth. But even when this "more-of-it" ideology is exposed as a failure, it remains difficult to dislodge as a solution because one can still appeal to its very nature. Not enough time, money, manpower, and so on were allowed so that positive results could be achieved. But if ever "more of it" were added, then success would result.

Finally, associated with the emphasis on utilization as an outcome is the failure to evaluate the effectiveness and appropriateness of medical practices and technology. This has led to health interventions that increase the cost of care but do not necessarily result in improvements in health status. Equal access to ineffective and inappropriate services will not improve the health of the population. While questions of access are important and legitimate, they are logically subsumed by considerations of effectiveness and appropriateness.

CONCLUSIONS

Without claiming they are definitive findings, and eschewing pretentions to an analysis as sophisticated as that of McKeown (1976a, 1976b; McKeown and Record, 1955, 1962; McKeown et al., 1972, 1975) for England and Wales, one can reasonably draw the following conclusions from the analysis presented in this chapter:

(1) That medical measures (both chemotherapeutic and prophylactic) appear to have contributed little to the overall decline in mortality in the United States since about 1900 — having in many instances been introduced several decades after a marked decline had already set in and having no detectable influence in most instances.

(2) That for those five conditions (influenza, pneumonia, diphtheria, whooping cough and poliomyelitis) for which the decline in mortality appears substantial after the point of intervention — and on the unlikely assumption that all of this decline is attributable to the intervention — at most 3.5 percent of the total decline in mortality since 1900 could be ascribed to medical measures introduced for the diseases considered here.

(3) From such data as are now available, and from all other analyses to date, there is not evidence for the commonly held view that medical measures (as distinct from public health) and the presence of medical services has detectably contributed to the observable decline in mortality that has occurred

in the United States (and many other countries) since about the turn of the century.

(4) With respect to morbidity, the evidence examined here indicates that there are no grounds for claiming or supposing that the health of the population — as measured by limitation in activity — is indeed improving to the extent that some believe is indicated by overall mortality trends. It is hoped that, with the continued collection of data over the next decades, this trend can be monitored more closely and a clearer picture can be obtained of the ongoing health status of the population.

(5) Despite more than a decade of major government interventions in medical care and considerable equalization of physician and hospital services utilization, persistent gaps remain in access to *quality* medical care, along with inequalities in health status.

Questions have been raised about our thesis (Levine et al., 1981) — that specific medical measures and the presence of medical care are not, in general, related to the modern decline in mortality or to improvements in the health of the population. It is therefore important to clarify what this argument does and does not hold. First, we are not arguing that medical care has no impact on the health status or quality of life of individuals. Second, this chapter should not be construed as removing support from a policy of equality of access to effective care. Rather, we are arguing that the burden of proof for the effectiveness of a particular medical intervention rests with the supporters of that measure. Furthermore, evaluation of these measures must take into account the potential for contradicting results among different outcome measures. For example, improvements in life expectancy may lead to reductions in the quality of life as a result of the side effects of treatment or increased years of disability. We can no longer afford the luxury of the prevailing ideology in Western liberal democracies that the simple addition of more financial resources to the medical care system will somehow result in better health.

POLICY IMPLICATIONS

The thesis we have reintroduced in this chapter — that medical measures were generally not responsible for most of the decline or changes in mortality and morbidity in the United States since 1900 — raises issues of the most strategic significance for researchers and health care legislators. Profound policy implications follow from either a confirmation or a rejection of the thesis. If one subscribes to the view that we are slowly but surely eliminating one disease after another by means of medical interventions, then there may be little commitment to social change and even resistance to some reordering of priorities in medical expenditures. If X is changing or disappearing primarily because of the presence of Y, then clearly Y should be left intact, or, preferably, be expanded. Its demonstrable contribution justifies its presence. However, if it can be shown convincingly, and on commonly accepted grounds, that the major part of the decline or changes in mortality and morbidity is unrelated to medical care activities, then some commitment to social change and a reordering of priorities may ensue. If the disappearance of X is rarely related to the presence of Y, or even occurs in the absence of Y, then clearly the expansion and even the presence of Y can be reasonably questioned. Its demonstrable ineffectiveness justifies some reappraisal of its significance and the wisdom of expanding it in its existing form.

In our view, and very briefly, social policy concerning the allocation of ever scarce resources must rest on the following premise:

GOVERNMENT SHOULD NOT SUPPORT THROUGH PUBLIC FUNDING FOR GENERAL PUBLIC USE, ANY SERVICE, PROCEDURE, OR TECHNOLOGY, THE EFFECTIVENESS OF WHICH HAS NOT BEEN, OR CANNOT BE, DEMONSTRATED.

Three criteria may be employed to determine whether or not a proposed procedure, service, or technology should be publicly funded. In order of logical importance, they are:

(1) *Effectiveness.* Whatever the intervention, it must first demonstrate some ability to alter beneficially the natural course of a clearly defined condition or set of conditions.

With few exceptions, acceptable evidence of an intervention's effectiveness can be established only through comparative experiments (RCTs) that are as free as possible of sources of bias. Then and only then can there be confidence that *the observed effect (if there is one) is actually the result of the intervention.* Such experiments must include at least the characteristics described by McKinlay (1981; see also Chalmers, 1981; Nyberg, 1974; Wulff, 1977; Lionel and Hexelheimer, 1970).

(2) *Cost Efficiency.* Where two or more proposed interventions of approximately equivalent effectiveness are available, that one should be preferred which involves the least cost.

(3) *Acceptability and Equity.* A proposed intervention, which is both effective and cost efficient is of no value unless it is socially acceptable and equally accessible to all the relevant sub-groups of the society into which it is being introduced.

Three caveats must be emphasized regarding any implementation of this general policy. First, there is no suggestion that these criteria, effectiveness, cost efficiency, social acceptability, and equality of access, should be applied only to innovations newly proposed for public funding. Clearly, innovations already ensconced in our publicly funded human services system must be subjected to the same scrutiny.

Second, there is no suggestion that the criteria proposed should be applied only to particular innovations, or only to those proposed by particular groups. Any intervention proposed for public funding (whether acupuncture, cardiothoracic surgery, chiropractic, social work, or transcendental meditation) should be subject to the same basic criteria.

Finally, it is no part of the present argument that ineffective services should necessarily be declared illegal, or removed from the human service marketplace (McKinlay, 1978). People should be free to purchase just about any human service desired — no matter how ineffective. What is questioned here is whether, through wasteful public expenditures and so forth, the rest of society should be required to pay for such prodigal purchases.

NOTES

1. Deaths in the category of chronic respiratory diseases (chronic bronchitis, asthma, emphysema, and other chronic obstructive lung diseases) could not be included in the group of chronic conditions because of insurmountable difficulties inherent in the many changes in disease classification and in the tabulation of statistics.

2. Barker and Rose (1976: 6) cite one study that compared the antemortem and autopsy diagnoses in 9,501 deaths occurring in 75 different hospitals. Apparently the overall frequency for a number of the diseases was very similar in the antemortem and postmortem diagnoses, despite many disagreements in individual patients. As an example they note that clinical diagnoses of carcinoma of the rectum were confirmed at autopsy in only 67 percent of cases, but the incorrect clinical diagnoses were balanced by an almost identical number of lesions diagnosed for the first time at autopsy.

3. Data from the two well-known Social Security Surveys were not included in this analysis as their estimates were, unfortunately, not directly comparable to those of the Health Interview Survey and presently cover a time span of only six years.

4. A "restricted activity day" is one in which a person cuts down on his or her usual activities for the whole of that day because of illness or injury. A person is "limited in activity" when he or she is limited in the amount or kind of some usual activity as a result of one or more chronic conditions. "Major activity" refers to the ability to work, keep house, engage in school or preschool activities, and the like. For more detailed information on these definitions, see Wilder (1976).

REFERENCES

Aday, L. and R. M. Anderson (1981) "Equity of access to medical care: a conceptual and empirical overview." *Medical Care* 12 (Supplement): 4-27.

Alderson, M. R. (1967) " Data on sickness absence in some recent publications of the Ministry of Pensions and National Insurance." *British Journal of Preventive and Social Medicine* 27: 1-6.

Alderson, M. (1976) *An Introduction to Epidemiology.* London: Macmillan.

Anderson, R., J. Kravits, and O. W. Anderson (1975) *Equity in Services.* Cambridge, MA: Ballinger.

Anderson, R., J. Lion, and O. W. Anderson (1976) *Two Decades of Health Services: Social Survey Trends in Use and Expenditure.* Cambridge, MA: Ballinger.

Barker, D. J. P. and G. Rose (1976) *Epidemiology in Medical Practice.* London: Churchill Livingstone.

Benham, L. and A. Benham (1976) "The impact of incremental medical services in health status, 1973-1980," pp. 97-103 in R. Anderson, J. Kravits, and O. W. Anderson (eds.) *Equity in Health Services.* Cambridge, MA: Ballinger.

Bolduan, C. F. (1930) *How to Protect Children from Diphtheria.* New York: New York City Health Department.

Bradford Hill, A. (1950) "Snow — an appreciation." *Proceedings of the Royal Society of Medicine* 48: 1008-1012.

Cannell, C. F., K. H. Marquis, and A. Laurent (1977) *A Summary of Studies of Interviewing Methodology.* National Center for Health Statistics, Vital and Health Statistics Series 2, No. 69. DHEW Publication (HRA) 77-1343. Washington, DC: U.S. Government Printing Office.

Chalmers, T. C. (1981) "The clinical trial." *Milbank Memorial Fund Quarterly* 59: 324-339.

Crichton, A. (1980) "Equality: a concept in Canadian Health care: from intention to reality of provision." *Social Science and Medicine* 14C: 243-257.

Davis, K. (1976) "Achievements and problems of Medicaid." *Public Health Reports* 91: 309-316.

Davis, K., M. Gold, and D. Makuc (1981) "Access to health care for the poor: does the gap remain?" *Annual Review of Public Health* 2: 159-182.

Diehr, P. K., W. C. Richardson, S. M. Shortell, and J. P. Lo Gerfo (1979) "Increased access to medical care: the impact on health." *Medical Care* 17: 989-999.

Donabedian, A., (1976) "Effects of Medicare and Medicaid on access to and quality of health care." *Public Health Reports* 91: 322-331.

Dubos, R. (1959) *Mirage of Health.* New York: Harper & Row.

Dubos, R. (1965) *Man Adapting.* New Haven, CT: Yale University Press.

Dubos, R. (1968) *Man, Medicine, and Environment.* London: Pall Mall.

Duff, R. (1976) "Patient care, the poor, and medical education," pp. 351-396 in J. Kosa and I. K. Zola (eds.) *Poverty and Health: A Sociological Analysis.* Cambridge, MA: Harvard University Press.

Dunn, H. L. and W. Shackley (1945) *Comparison of Cause of Death Assignments by the 1929 and 1938 Revisions of the International List: Deaths in the United States, 1940 Vital Statistics.* Special Reports 19: 153-277, 1944. Washington, DC: U.S. Department of Commerce, Bureau of the Census.

Fuchs, V. R. (1974) *Who Shall live?* New York: Basic Books.

Goodrich, C. H., M. C. Olendzki, and G. G. Reader (1970) *Welfare Medicine: An Experiment.* Cambridge, MA: Harvard University Press.

Griffith, T. (1967) *Population Problems in the Age of Malthus.* London: Frank Cass.

Gutmann, A. (1981) "For and against: equal access to health care." *Milbank Memorial Fund Quarterly* 59: 542-560.

Habakkuk, H. J. (1953) "English population in the eighteenth century." *Economic History Review* 6.

Hart, J. T. (1971) "The inverse care law." *Lancet* (February 27): 405-412.

Inman, W. H. W. and A. M. Adelstein (1969) "Rise and fall of asthma mortality in England and Wales, in relation to use of pressurized aerosols." *Lancet* 2: 279-285.

Kass, E. H. (1971) "Infectious diseases and social change." *Journal of Infectious Diseases* 123: 110-115.

Krugger, D. E. (1966) "New enumerators for old denominators — multiple causes of death," pp. 431-443 in W. Haenszel (ed.) *Epidemiological Approaches to the Study of Cancer and Other Chronic Diseases.* Washington, DC: U.S. Government Printing Office.

Lee, W. W. (1931) "Diphtheria immunization in Philadelphia and New York City." *Journal of Preventive Medicine* 5: 211-220.

Levine, S., J. Feldman, and J. Elinson (1981) "Does medical care do any good?" Paper presented at the meetings of the American Sociological Association, Toronto.

Lilienfeld, A. M. (1976) *Foundations of Epidemiology.* New York: Oxford University Press.

Lionel, N. D. W. and A. Hexelheimer (1970) "Assessing reports of therapeutic trials." *British Medical Journal* 3: 637-640.

Luft, H. (1978) *Poverty and Health: Economic Causes and Consequences of Health Problems.* Cambridge, MA: Ballinger.

Magill, T. P. (1955) "The immunologist and the evil spirits." *Journal of Immunology* 74: 1-8.

McKeown, T. (1976a) *The Modern Rise of Population.* London: Edward Arnold.

McKeown, T. (1976b) *The Role of Medicine: Dream, Mirage of Nemesis?* London: Nuffield Provincial Hospitals Trust.

McKeown, T. and R. G. Record (1962) "Reasons for the decline in mortality in England and Wales during the nineteenth century." *Population Studies* 16: 94-122.

McKeown, T., R. G. Brown, and R. G. Record (1972) "An interpretation of the modern rise of population in Europe." *Population Studies* 26: 345-382.

McKeown, T., R. G. Record, and R. D. Turner (1975) "An interpretation of the decline of mortality in England and Wales during the twentieth century." *Population Studies* 29: 391-422.

McKinlay, J. B. (1977) "On the medical-industrial complex." *American Medical News* (April 11): 26.

McKinlay, J. B. (1978) "The Limits of Human Services." *Social Policy* 8: 29-36.

McKinlay, J. B. and S. M. McKinlay (1977) "The questionable contribution of medical measures to the decline of mortality in the United States in the twentieth century." *Milbank Memorial Fund Quarterly* 55: 405-428.

McKinlay, J. B. and S. M. McKinlay (forthcoming) "A refutation of the thesis that the health of the nation is improving."

McKinlay, S. M. (1981) "Experimentation in human populations." *Milbank Memorial Fund Quarterly* 59: 308-323.

Moriyama, I. M., W. S. Bau, W. M. Haenszel, and B. F. Mattison (1958) "Inquiry into diagnostic evidence supporting medical certifications of death." *American Journal of Public Health* 48: 1376-1387.

Moriyama, I. M. and S. O. Gustavus (1972) *Cohort Mortality and Survivorship: United States Death-Registration States, 1900-1968.* National Center for Health Statistics Series 3, No. 16. Washington, DC: U.S. Government Printing Office.

Morris, J. N. (1965) "Sickness absence — return to work?" *Proceedings of the Royal Society of Medicine* 58: 821-825.

Nyberg, G. (1974) "Assessment of papers of clinical trials." *Medical Journal of Australia:* 381.

O'Connor, J. (1973) *The Fiscal Crisis of the State.* New York: St. Martin's.

Pfizer, C. and Company (1953) *The Pneumonias, Management with Antibiotic Therapy.* Brooklyn: Author.

Reid, O. D. and G. A. Rose (1964) "Assessing the comparability of mortality statistics." *British Medical Journal* 2: 1437-1439.

Rein, M. (1969a) "Social class and the health service." *New Society* 14: 807-810.

Rein, M. (1969b) "Social class and the utilization of medical care services." *Hospitals* 43: 43-54.

Schneyer, S., J. S. Landenfeld, and F. H. Sanifer (1980) "Biomedical research and illness: 1900-1979." *Milbank Memorial Fund Quarterly* 59.

Schroeder, S. (1981) "Editorial: national health insurance — always just around the corner?" *American Journal of Public Health* 71: 1101-1103.

Sidel, V. W. and R. Sidel (1977) *A Healthy State: An International Perspective on the Crisis in United States Medical Care.* New York: Pantheon.

Somers, A. R. and H. M. Somers (1977) *Health and Health Care: Policies and Perspectives.* Germantown, MD: Aspen Systems.

Stoeckle, J. S. (1976) "The reorganization of practice in the community," pp. 351-396 in J. Kosa and I. K. Zola (eds.) *Poverty and Health: A Sociological Analysis.* Cambridge, MA: Harvard University Press.

Sullivan, D. F. (1971) "A single index of mortality and morbidity." *HSMHA Health Reports* 86: 347-354.

Syme, S. L. and L. F. Berkman (1981) "Social cass, susceptibility and sickness," pp. 35-36 in P. Conrad and R. Kern (eds.) *The Sociology of Health and Illness: Critical Perspectives.* New York: St. Martin's.

Taylor, P. J. (1971) "Some emotional hazards of work." *Public Health* 85: 298-302.

Titmuss, R. M. (1968) *Commitment to Welfare.* London: Allen & Unwin.

Townsend, P. (1974) "Inequality and the health services." *Lancet* (June 15): 1186.

U.S. Bureau of the Census (1906) *Mortality Statistics 1900-1904.* Washington, DC: U.S. Government Printing Office.

U.S. Department of Health, Education and Welfare (1964) *The Change in Mortality Trends in the United States.* National Center for Health Statistics Series 3, No. 1. Washington, DC: U.S. Government Printing Office.

U.S. Department of Health, Education and Welfare (1979) *Health, United States: 1979.* Publication (PHS) 80-1232. Washington, DC: U.S. Government Printing Office.

U.S. Department of Health, Education and Welfare (1980) *Health, United States: 1980.* Publication (PHS) 81-1232. Washington, DC: U.S. Government Printing Office.

U.S. Government (1966) *Social Security Survey.* Washington, DC: U.S. Government Printing Office.

U.S. Government (1972) *Social Security Survey.* Washington, DC: U.S. Government Printing Office.

Weeks, A. H. (1977) "Income and disease — the pathology of poverty," pp. 53-65 in L. Corey, M. F. Epstein, and S. E. Saltman (eds.) *Medicine in a Changing Society*. St. Louis: Mosby.

Weinstein, L. (1974) "Infectious disease: retrospect and reminiscence." *Journal of Infectious Diseases* 129: 480-492.

Wilder, C. S. (1976) *Health Characteristics of Persons with Chronic Activity Limitation, United States, 1974*. National Center for Health Statistics, Vital and Health Statistics Series 10, No. 112, DHEW Publication (HRA) 77-1539. Washington, DC: U.S. Government Printing Office.

Wilson, G. S. and A. A. Miles (1946) *Topley and Wilson's Principles of Bacteriology and Immunity*. Wiltimore: Williams & Wilkins.

Wilson, R. and E. White (1977) "Changes in morbidity, disability and utilization differentials between the poor and non-poor: data from the health interview survey: 1964 and 1973." *Medical Care* 15: 636-646.

Wulff, H. R. (1977) "Check list for assessment of controlled therapeutic trials." *Acta Neurologica Scandinavia* 60 (Supplement): 79-80.

Wyszewianski, L. and A. Donabedian (1981) "Equity in the distribution of care." *Medical Care* 12 (Supplement): 28-56.

5

The City's Weakest Dependents:
The Mentally Ill and the Elderly

ANN LENNARSON GREER
SCOTT GREER
TOM ANDERSON

☐ NOTHING SO EXEMPLIFIES the nostalgic character of the American political folk culture as the recurrent suggestion that the frail elderly be returned to the care of their children. The problem of such dependent populations, unemployed and unemployable, sick and intrinsically powerless, is an eluctable product of the increasing societal scale and the growth of cities as major population concentrations. As we concentrate populations, we concentrate misery and the visibility of misery. Reports of the mass media are supplemented by knowledge at first hand. We learn there are thousands of people without homes who haunt the subways, streets, and sewers of our great cities; we see them pilfering garbage cans on the sidewalk by Fifth Avenue. We know there are the dangerous deviants and the pitiful deviants. Not so much in their incidence (for they occur in all populations) but in their concentration, they are symbols of the dark side of the city, standing for the precariousness of human existence, insanity, illness, old age, and death.

These are the city's weakest dependents. In all societies they present the urban dweller with examples of misery, for the city attracts them (they "drift" there), and within the city they are concentrated at the center, where they beg and sometimes live. In a highly urbanized society, they are a responsibility of and a burden to the city's more able population; in a modern, technological society, they are also apt to be "medicalized." That is, they are seen as "without fault," hence deserving of care, by the population in general and especially by the professional medical workers under whose jurisdiction they fall.

In this chapter we will discuss the general problem of care for the city's dependent populations in the past, emphasizing the West and the United States. From this discussion we then move to a consideration of two populations who have been medicalized and who use a large and increasing proportion of the health services of the city, particularly the older central city. The mentally ill and mentally retarded are discussed within the general framework of "cycles of care," from local community care to institutionalization and back. This historical pattern is discussed as it is implicated in the present effort at caring for the mentally ill in the United States. In this discussion we rely on the literature and on our empirical studies of community mental health centers (Greer and Greer, 1980). We then turn to another population, the frail elderly, and discuss their care in similar terms, within a similar framework, relying on both the literature and our investigation of care for the elderly in Wisconsin. Frequently, of course, these populations overlap, and the goal of "returning them to the community" is often attractive; to what extent this is grounded in possibility and to what extent in fantasy we will try to determine. One thing is certain: These populations will continue to exist and increase in our cities. As one observer (Friedman, 1982) described the problem of hospital "dumping" of indigent patients, "The poor are always with some of us."

HEALTH AND HANDICAPS
IN URBAN HISTORY

By the social bookkeepers' categories, we are discussing the poor who are also old and sick or insane: the mentally retarded, alcoholic, hopelessly addicted, blind, crippled, or, sometimes, children without adult protectors. What they have in common is weakness: poverty, lack of utility for others, political powerlessness. They are wards of the state, if of anyone. The generosity of contemporary large-scale and affluent societies toward such people, uneven as it is, is rare in human history. For the most part, such populations have simply gone to the wall. With a few shining exceptions, in primitive societies the nonfunctional, who also lacked power, wealth, or claims on kin, were treated indifferently at best (Eisenstadt, 1956; de Beauvoir, 1972). In folk societies, subsistence agricultural communities controlled and exploited by an urban-centered order, the pattern was not dissimilar. Effective power in the political order or the economy, most often through ownership of land, guaranteed support. It earned respect and care through old age and illness. But as Lear discovered, the old relinquished their claims on power at their peril (de Beauvoir, 1972). The old poor, and handicapped persons in general, were first the responsibility of their families and beyond this, whatever community they ended up living in. It was often a very weak support.

This condition is not surprising when we remember how poor most past societies generally were. The urban revolution, a process that can be seen as a spatial division of labor, control, and wealth, did not automatically lead to a dependable affluence. The cities of northern Italy during the sixteenth century were among the wealthiest of the time, yet their people suffered recurrent famine (Braudel, 1976). When such famine was due to crop failure in the hinterland, on whose products most city-states were heavily dependent, the walls of the cities were shut against the starving peasants who flocked toward the city's granaries. The city was a concentration of wealth within a poor society, itself dependent on long, expensive, and precarious lines of communication and transportation.

These microcosms of urbanism forshadowed much of the shape of our cities today. The city, a part of a spatial division of labor, was based on an intricate division of labor internally.

This included the market mechanism, from financiers and merchants to stevedores and teamsters. It also included the court, from prince to tax collectors and watchmen (and swarms of parasites). Artisan industry became increasingly influential in such cities as Florence, with gradations from guild master to apprentice and casual laborer. Finally, the church, with its various organizational segments, was a major source of occupation. From such differentiation emerged stratification systems, with the dominant order (whether church, state or business) at the top, and closely callibrated differences within each order. Such organization multiplied pseudo-species and increased social distance between them; those who did not belong were, so to speak, socially deracinated (Erikson, 1963).

At the same time, the movement from village to city tended to leach out older claims to support by unproductive dependents. The family, no longer cohering around land as a place of work and residence, was less able and willing to support them, even when it also lived in the city. The urban community beyond the family was relatively alien and disinterested. Whatever responsibility was assumed, was assumed in the name of Christian charity; to give alms to the poor counted toward salvation. Such alms were somewhat centralized in the "hostels" of religious orders, where the beggar retreated when the streets were unproductive. There the pauper, the sick, the insane, and the generally homeless and unprotected were sheltered (Rosen, 1963; Glaser, 1968).

With the increasing surplus brought about by improved agriculture and increasing trade, more would be done for the weakest dependents. In England, from which so many of our own institutions and attitudes derive, the care of these populations was always a responsibility of kin and local government. But one must ask at this point: Where do we derive the social purpose for funding such dependency?

We know of no rigorous, empirically based, explanation; perhaps there is no single cause of such societal norms. With respect to the old, they are auguries for the young who support them as a kind of insurance for their own phase of age and weakness. Something akin to "Honor they father and mother . . . that thy days may be long" is found in all the great world religions. It is a sentiment that combines altruism with self-interest (Cottrell, 1955). At the same time the principle is extended to the other dependents, in part because, with increasing societal surplus, they stay alive more easily, and a great many families (from royalty to beggars) have members who are crippled, blind, mentally retarded, or otherwise handicapped. There is often loyalty and identification with these folk by the kinsmen. There seem to be many discrete causes for what W.J. Goode (1973) has called "the protection of the inept."

Because of such sentiments, part altruism and part self-interest, the care of the weakest dependents still fell mostly upon kinfolk. But in the urban milieu, the public almshouses, hospitals and hostels were developed as partial solutions to the various problem-presenting populations. With increasing population and erratic economic conditions, English society had a substantial number of people, handicapped and not, who drifted in and out of poverty. With the Elizabethan Poor Law of 1601, the national state first assumed responsibility for the direction and nature of support for the helpless. It was, in line with the English political system, decentralized to the local area for support and administration. Differentiating dependents as children, able-bodied and "impotent," the Poor Law provided care in the form of "outdoor" and "indoor" relief. Those still having a place to live and capable of caring for themselves received an outdoor stipend, while the homeless and incompetent received indoor assistance.

Thus, at the beginning of the Industrial Revolution, the task of caring for the helpless was shared by public almshouses, jails, insane asylums, and hospitals. The categories of clients did not necessarily match the stated purpose of the sheltering organization; the boundaries were not clear between cases, and many individuals suffered from several of the defects associated with dependency. Paupers, thieves, the mentally retarded and mad, along with those ill of body, might be housed with homeless children and abandoned elders.

As the demand for labor inreased relative to the population, preventing unemployment, and as the increasing societal surplus allowed for support of the unemployed, *poverty* as a "problem" was mooted (McNeil, 1982). Voluntary charity from the private agencies that had grown with urbanization supplemented the dole granted by local authorities. These provided a minimal charity for the deserving poor — those who could not help themselves — and attempted to make life as difficult as possible for the others. Volunteerism in the egalitarian form of self-help created labor unions, and eventually the territorial state underwrote both efforts. Thus the only remnants of true paupers were the "undeserving poor," who could work but would not do so under the terms available to them. Today we use the terms "hard-core unemployed," "welfare chiselers," and so forth to refer to the undeserving poor. The "welfare state" ideally provides full employment for able-bodied people who want to work (unless they have other claims on surplus through property or kin), while unemployment compensation supports them through periods of less demand for their labor.

The elimination of pauperism from the charges on the local community left the disabled — those who could not work or, at an extreme, even exist independently in ordinary society. The physically disabled include the blind, the crippled, the ill, and often the old. The mentally disabled include those who are diagnosed as mentally ill and those who are mentally retarded; they are also often aged. In both types of disability, a large proportion of the victims are judged to be permanently disabled; they are called the "chronics." Thus, each type of disability can be seen as ranging from the curable, who could with proper attention be returned to independent social life, and the incurable, who are permanent dependents. Between lies a gray area that has been the site of continuous ideological and political conflict for many years, for the distinction between the two types is unclear in many disabilities, and varies in ambiguity among individuals.

The United States in its early days was a prosperous, mostly rural, mostly agriculturally based society. Its population was young, land was new to cultivation and had not yet been exhausted, and new land lay readily at hand. The country was labor-poor and land-rich; thus, subsistence was usually not a problem for either the farmer or the hired laborer. Most farms were nearly self-supporting, and cash crops were exported to cities. All in all, as the country sowed seeds of industrialization in scattered application of nonhuman energy sources (falling water, canals, steam power, railroads), it was a nation of small towns and open country neighborhoods.

Thus, the origins of American support for the dependent were based on the English pattern of local responsibility, with towns and rural counties caring for the sick and helpless. Furthermore, care was local within a profoundly "federal" system: each state had its own history, its own governmental structure. The township was basic to New England, but as one went west and south into more sparsely populated farm country, the county became the dominant local form of government (Webb, 1931; Boorstin, 1965). In both kinds of political community the public care of the dependents was usually the county farm, or poor farm. This institution was supposed to provide security, protection, healthy existence with exercise, and economic self-sufficiency. (Indeed, such farming could result in profitable agriculture.)

In the city, matters were otherwise. The number and diversity of dependents was so great, their concentration so obvious, that the government devised various catchment pools. As in urban England, jails, hospitals, madhouses, and hostels were crowded with the unwanted human surplus. A recent study of Philadelphia General Hospital, the oldest civic institution of the sort in the country, details the mix of social dependents the hospital accepted. Police vans and ambulances, relatives and courts, continually dumped clients on the doorstep of the institution. While some of the most brilliant modern medicine was developed in parts of the hospital, the "back wards" contained the abandoned aged, poor, and chronic mental patients. Ominously, when the hospital closed in 1977, the back wards were still there, anomalies in a modern general hospital, but not in contemporary urban society (Rosenberg, 1982).

In the United States, Dorothea Dix first called national attention to the haphazard treatment (often cruel, more often callous) of the insane. Her crusade for providing humane asylum to the afflicted, based on the propositions that they *were* medically ill and that they could be cured, had a major influence on the social definition of the mentally ill. By separating them physically, in state hospitals, and categorically, as subject to medical treatment, she removed the justification for treating them as an undifferentiated part of the larger mass of the helpless.

The profession of psychiatry emerged from the job of running these hospitals and the task of learning more about the nature of mental illness and its cure. With what today might be called behavioral therapy and milieu therapy, positive results were obtained. Small institutions with atmospheres of support and personally responsible physicians seemed to be therapeutic. Without these conditions, however, recovery and faith in recovery decline. As the century wore on, the increasing number of "incurably insane" in large, poorly staffed institutions tended to produce therapeutic fatalism in those responsible for care, from legislators to physicians. Medical theory tended to follow, and speculative constructs of "brain lesions" and the like were evoked as explanations.

With the growth of industrial America, immigrants poured into the cities and into the institutions. Most were new to urban life, many to American society and the English language. Urban growth led to the rapid growth of huge state mental hospitals, which flooded the small, "humane care" institutions for which Dorothea Dix had crusaded. It was popular in the nineteenth century to blame the increased numbers of the dependent population on the inferior nature of the new immigrants. Dependency more likely increased because of (1) the sheer growth of the urban population and therefore the disabled and (2) the concentration of this increase in the cities, where they were visible.

Under these circumstances, public provision for the helpless was less than inadequate. The public welfare agencies we know were being slowly and painfully invented. Some of the surrogates, and sources of innovations, were private charities that, on the English pattern, supplemented governmentally based efforts. The current mixture of public and private resources and control has its roots deep in the period when mass public support for the dependent in large-scale American society was emerging. Public agencies using private volunteers who serve without pay and private agencies funded by the public but controlled by voluntary citizen boards are the heirs of this support system in the United States. Dorothea Dix is one in a long line of committed and talented private persons who helped shape care for the helpless; we find her like today in volunteers creating child-guidance centers, shelters for the homeless aged, and community mental health centers (Greer and Greer, 1980; Greer, 1982).

Nevertheless, the Great Depression of the 1930s established the principle of public responsibility for the helpless, and particularly the role of the national state. In the atmosphere of crisis prevailing in the decade, new subjects of concern developed and old ones received more care than before. This process resulted in a rapid growth of the labor force servicing dependents and in the development of large-scale bureaucracies at the local, state, and federal levels to order that labor force. Then, in the increasingly affluent society after World War II, more *kinds* of problems were identified, and welfare work in turn became more differentiated.

With differentiation came specialization and eventually professionalization. The model was most likely medicine, in which the twentieth century has seen a dazzling array of specialties created, many of them responses to new science and technology (Stevens, 1971). Many others were a response to the division of labor. Here are people to be cared for; someone must care for them. In the treatment of the aged and mentally incompetent dependent populations, this was the norm. The physical ailments they shared with the "normal" could usually be treated by M. D.s, but their specific complaints and ineptness, immobility, destructive disease, and the like remained recalcitrant to appropriate treatments. It did not simplify matters that each ailment was open to speculative inquiry as to whether its cause was organic or behavioral or some combination of both.

Since some are thought to be mentally ill, they are so classified by the agencies responsible for their welfare and that of the public. Such classification leads to the elaboration of psychiatry and "abnormal" psychology. The congenitally retarded are subjects for various behavioral therapies. The aged fit many specialties, and finally a superspecialty, geriatrics, or the even broader gerontology. Other specialists lack complex titles but are experts in alcoholism, drug abuse, senility, paraplegia, the blind, and so on through the list of human frailties outside those defined as strictly medical. Finally there are those who specialize not in an ailment but in a procedure. Here we find many varieties of counselors — recreational therapists, group therapists, and so on, *ad infinitum*. Such specialties often owe more to perceived need and practice than to science (Bucher and Strauss, 1961; Schatzman and Strauss, 1966)

The roles these specialists play are usually in formal organizations and bureaucracies. They are chosen according to their presumed education and experience for the role; they identify with it, and become responsible for the clients classified as appropriate to their job. But the professionalization of purveyors, practitioners, and advocates of different types of dependency tends to seal the individual client in a cell of bureaucratic order. Unfortunately, his problems often include multiple domains of care — he is old, poor, ill, mentally undependable, and physically incontinent. Most obviously, he is often beyond cure, and all that can be done is care. He is a net cost to the society.

While America places great emphasis on the treatment and cure of acute illness, it lacks concern for the multiplying problem of the noncurable. Progressive behavioral handicaps caused by physical degeneration and mental malfunctioning result in a large and growing dependent population (Wiener et al., 1982). These are the predictable demographic consequences of improving living conditions (Leavitt, 1982). Healthier cities produce larger cohorts who are subject to the hazards of aging. They also protect, however ineptly, the mentally incompetent, allowing their survival. They are the stepchildren of the medical culture; its practitioners continue, hoping for cure, grudgingly caring.

THE MENTALLY ILL IN
THE COMMUNITY

The state system of institutional care that evolved in America during the nineteenth century to care for the mentally disabled must fairly be characterized as reflecting two thrusts in the society of the time. The humanitarian movement spearheaded by Dorothea Dix preached both the hope of cure and the refuge of asylum (Grob, 1980; Rothman, 1971; Gruenberg and Archer, 1979). The movement's humanitarian motivation arose, of course, in response to its members' perception that ordinary motivation arose, of course, in response to its members' perception that ordinary treatment of the mentally incapable was neither therapeutic nor compassionate. Reformers urged the creation of state asylums so that sympathy, treatment, and enlightened guardianship could replace the misunderstanding, degradation, and punishment the mentally ill suffered in local community jails and poorhouses.

In their first annual report, the trustees of one state asylum described this goal. The institution, they stated, was

> for those people who have neither home nor friends, and who are without the means financially or capacity intellectually to provide for themselves, with intellect shattered, minds darkened, living amid delusions, a constant prey to unrest, haunted by unreal fancies and wild imagining. They now have

in their sore misfortune a safe refuge, kindly care, constant
watching, and are as comfortable as their circumstances will
allow. This is a result over which every human and Christian
citizen of the state will rejoice [Bassuk and Gerson, 1978: 53].

Clearly these institutions met an existing and growing demand. Large-scale immigration increased the numbers of
homeless, friendless persons mentally or socially unable to
cope and who were, we must emphasize, unwanted (Mechanic,
1969: 73-77). During the later nineteenth and early twentieth
centuries, the institutions grew ever larger and more ethnic,
containing large numbers of immigrants who, economically
marginal upon arrival, unskilled and uneducated, were especially likely to become wards of the state when illness struck.

Because the immigrant patients came disproportionately
from the growing industrial cities, they were magnets for the
ambivalence of the native-born to the changes occurring in
their society. Grob (1980) notes that the arrival of the Irish in
Massachusetts signaled the change of that state's economy
from one oriented to small-scale skilled enterprises to one of
large-scale unskilled industries. Increasingly, the "argument
was heard that the public burden of supporting foreign-born
paupers, including the insane, was fast becoming unbearable"
(Grob, 1980: 38) Theories such as phrenology, which linked
mental incapacity to appearance, became popular, as did
theories that derived mental illness from hereditary predisposi-
tion or (foreign) living habits (Grob, 1980: 33-35). Although
proponents of deinstitutionalization existed during this period
to argue the harmlessness of most patients, their appeals received little attention (Scull, 1977). Like the madhouse of earlier days, the insane asylum was once again viewed as existing
not to protect the deranged individual, but to protect the community from the individual.

Under the pressures of large numbers, rising costs,
therapeutic nihilism, and public antipathy, hospitalization became increasingly custodial in its orientation. Although the
medical model was retained, untrained and low-paid attendants
substituted for psychiatrists almost all of the time. Physical
restraint and solitary confinement were increasingly the

mechanisms of control (Grob, 1980: 35). Even the emergence of psychiatry as a profession during these years may have had deleterious consequences through removal of the interested laypeople who had in the earlier epoch provided optimism and humanitarianism.

Not all of the inmates were foreigners. The hospitals continued to serve their enduring function of absorbing the native-born paupers and "others who were disruptive of the social order or burdensome to their families" (Morrisey et al., 1980: 291). Interviews conducted nationwide by two of the authors yielded repetitive accounts of state hospital histories.[1] Although sharing a presumptive focus on mental illness and a commitment power to back it up, twentieth-century mental hospitals were still used as almshouses receiving, segregating, and controlling not only the mentally ill but the rejected helpless in general. A representative of the Mental Health Association in Alabama told how a hospital built for 500 came to house and minimally to maintain a population of 1,600:

> What would happen would be when, say, family members decided they didn't want Aunt Mary around anymore, probably after managing to get away from Aunt Mary what few pennies she might have saved, they would drop her off at Searcy Hospital, probably with no clothes, just dump her off on the grounds. The hospital wouldn't have any address for the family members; they probably drove right from the hospital on to California or Detroit [Interview, 1978].

A representative of the hospital elaborated:

> [We didn't have] good guidelines for admission. It used to be a . . . man could say that his wife was driving him crazy, get a doctor to agree with him, and commit her [Interview, 1978].

Mechanic (1969: 75) cites a study showing that in New York state in 1940, 31 percent of all admissions to mental hospitals were persons diagnosed as senile or suffering from cerebral arteriosclerosis. He cites related studies that documented a relationship between economic fluctuations and admissions to psychiatric hospitals, concluding that "economic and social instability produced large numbers of persons in need of care, and the mental hospital in the absence of other alternatives assumed this function."

The populations of such institutions upon mid-twentieth-century examination were easily found to contain many persons whom citizens' groups, civil rights lawyers, and courts saw as wrongfully detained, often for life. In several important cases, beginning in the 1950s, U.S. courts ruled that involuntary incarceration in nontherapeutic mental hospitals was justified only for patients who were both documentably ill and dangerous. The courts instructed that, for all but this narrowly commitable group, treatment should be in the "least restrictive setting," that is, in community institutions.

The enthusiasm for community care that underlay these decisions was grounded in several intellectual and social developments. One reflected the public health movement of the early twentieth century, which had advanced the cause of "mental hygiene." Although those who campaigned on behalf of mental hygiene were concerned with the deteriorated conditions in the mental hospitals, their strategy for change was within the community, where they advocated prevention, early detection, and community treatment of mental disorders. The principal concretization of their efforts was the establishment in communities across the country of child guidance clinics focused on disorders of childhood and adolescence, family problems, and parental education (Gruenberg and Archer, 1979; Mechanic, 1969; Morrisey et al., 1980).

The introduction and rapid growth of psychotherapy as a treatment in the ensuing period furthered the development of community mental care. Large numbers of trained psychotherapists entered the United States in the 1930s as refugees; they established lucrative office practices and captured medical school posts. Their arrival and influence "put the office practice of psychiatry on a commercial basis" and "laid the foundation for psychiatric careers unconnected with the state mental hospitals" (Gruenberg and Archer, 1979: 489).

The positive events that provided the immediate impetus in the 1950s and 1960s for deinstitutionalization and led to specific action were several. Within psychiatry there was renewed therapeutic optimism stemming from experimentation with new psychosocial techniques such as milieu therapy, therapeutic communities, and "open hospitals." These strategies humanized, decentralized, and reduced restrictions in hospital care and stimulated proposals for decentralized hospital programs (Gruenberg and Archer, 1979; Myerson, 1980; Morrisey et al., 1980). They were augmented by the introduction in the late 1950s of psychoactive drugs, which allowed stabilization of formerly unpredictable patients, making them tractable citizens of their institutions and perhaps candidates for life in the community. One psychiatrist recalls the introduction of drugs this way:

> The effectiveness of modern psychotropic drugs came home to me a month or so back when I found by accident, in a cupboard at one of the local psychiatric hospitals, a former register of mechanical restraint and seclusion, which at one time had to be made out for every patient secluded (restraint was never employed) and had to be countersigned by myself as medical superintendent. I had completely forgotten how disturbed was the behavior of so many patients, those years ago, who had from time to time to be secluded from their fellow patients because of their aggressive conduct. The picture now has completely changed, a change due largely to the phenothiazines. It is true that more is initially claimed for new methods of treatment than the subsequent results confirm, and it has long been observed that new methods of treatment throughout the last fifty years and longer came in with a 90 percent recovery rate on first introduction, but eventually showed no more than the usual 30 percent remission [Bowen, 1979: 544].

Renewed therapeutic optimism dovetailed with harsh criticism of institutional care. The courts that ordered patients to be released from institutions were influenced not only by hopefulness for community care but by sociological studies documenting the debilitating effects on confined persons of institutional life per se (e.g., Goffman, 1961) and the degree to which the labeling of an individual as sick or afflicted can become a self-fulfilling prophecy (e.g., Becker, 1964). The result appears to have been, as Jones (1979: 560) observes, a point of view that "whatever the dangers and shortcomings of life in the community, what the community was likely to do to expatients could not possibly be worse than [institutionalization]."

Thus, state governments found themselves bearing the burden of the helpless but without any applause for their efforts. Ever escalating institutional costs met ever escalating criticism. The states, which had entered the picture through a humanitarian drive to rescue the mentally disabled from the harsh hand of local government, had become the villains. Eager to dispose of this economic and political liability, the states eagerly embraced the judicial decisions that permitted them to discharge mental patients and refuse the social residual they had previously accepted. As one state hospital official told us,

> This is the thing that makes it hard to understand. Before the court case we were trying to get rid of these patients. The families could say, "We don't want them." We had no clout to make anybody take patients that we thought didn't belong here. Now we have the court behind us. The court says "you will take these patients" [Interview, 1978].

Now the courts required it and, as we shall discuss, the federal government subsidized it. Mental patients were discharged to settings judged less restrictive. Censuses of the state mental hospital began to decline precipitously around 1955, eight years before the federal government enunciated its program of community mental care and provided resources to support community services (U.S. Comptroller General, 1977; Scull, 1977; Rose, 1979).

THE FEDERAL INITIATIVE: COMMUNITY MENTAL HEALTH CENTERS

When, in 1955, the U.S. Congress authorized the Joint Commission on Mental Illness and Health to study and make recommendations for federal action in the area of mental care, there was broad consensus on the desirability of deinstitutionalization of state mental patients. The commission's members, however, were divided between two broad approaches to the accomplishment of this goal. One group favored the "open hospital" model, wherein the staffs of renovated and reorganized state hospitals would arrange for the early discharge of patients, supervise their aftercare in the community, and oversee their rehospitalization in times of crisis (Gruenberg and Archer, 1979; Myerson, 1980; Morrisey and Goldman, 1980). The other group advocated a community services model wherein a diverse array of professionals and community groups could assemble services appropriate to community needs, with the objectives of preventing mental illness, averting long-term hospitalization through preemptive treatment, and rehabilitating the severely ill (Gruenberg and Archer, 1979; Mechanic, 1969; Bassuk and Gerson, 1978). Differences between the two approaches concerned the degree to which chronic hospital patients were the focus of attention, the extent to which the hospital and the medical care model would serve to organize future care, and the relative optimism for cure or rehabilitation that they offered.

Although the commission's report, *Action for Mental Health,* which was published in 1961, straddled the differences between these approaches, implementation required that choices be made. It was the more radical and more optimistic community model that captured the imagination of President Kennedy and served to shape the Community Mental Health Centers Act of 1963 (Gruenberg and Archer, 1979; Mechanic, 1969). The 1963 act authorized the construction of mental health centers in 1,500 community "catchment areas" of 75,000-200,000 population. The ultimate goal was to blanket the country with 1,500 of these centers. To be designated, applicant groups were required to provide five essential preventive and rehabilitative mental care services (acute inpatient care, outpatient care, emergency treatment, partial hospitalization, and consultation and education). The centers (CMHCs) were accountable to the National Institute of Mental Health (NIMH), near Washington, D.C.

Kennedy's optimism that community-based services could ultimately replace institutional care was evident in his message announcing the CMHC program. In it he described the future role of the state hospitals as follows:

> Until the community mental health center program develops fully, it is imperative that the quality of care in existing State mental institutions can be improved . . . [for] *many such institutions can perform a valuable transitional role* [quoted by Gruenberg and Archer, 1979: 494; emphasis added].

Needless to say, this left-handed encouragement did not lead states to invest in their mental institutions. In the years to follow, states accelerated their discharge of all but the sickest patients. By 1975, hospital censuses had been reduced to one-third their 1955 levels (Scull, 1977; Bassuk and Gerson, 1978; U.S. Comptroller General, 1977). The receptiveness of the community systems to the patients was, however, often minimal.

Several aspects of the CMHC programs are important to understanding their development as independent systems of care largely unrelated to the old state hospitals and their clients. As noted, the programs' very conception minimized the problem of discharged state hospital patients. The structure, governance, and funding of the centers were focused not on the chronically ill patients being discharged from mental institutions, but on the provision of services to patients who could be successfully cured or rehabilitated.

It is noteworthy that there was no required relationship to the chronic care state hospitals or the patients discharged from them — either to receive or keep track of discharged patients — and no need to demonstrate the complex *integration* of services that such patients require. An acceptable model for CMHC organization, for example, was for the required services to be provided by a consortia of providers. It is possible that there are funding consortia that have managed to integrate patient records and referrals, but we know of none. It is clearly much easier to achieve funding coalitions than to integrate patient care services. The medical aspects of mental care were themselves deemphasized. The program discouraged medical sponsors such as hospitals from submitting grant applications by requiring centers to have freestanding governing boards.

This provision also precluded application by existing government units. This exclusion expressed the federal government's suspicion of local government and probably made testy relationships with local units inevitable. It created difficult problems in states such as Wisconsin, where county governments had an eighty-year history of responsibility for their mentally ill, and in other states, which responded to the deinstitutionalization movement of 1960s and 1970s by *assigning* the responsibility for the mentally ill to local government jurisdictions.

Not surprisingly, the nonprofit citizen boards that fit the federal government's idea of nonmedical, nongovernmental sponsorship were typically those running child guidance clinics or related private agencies dedicated to sociopsychological approaches to mental trauma. Although of necessity the National Institute of Mental Health made many compromises with hospitals and local governments (without which the program could not have been implemented at all in whole regions of the country), preferred applicants were private-sector bodies whose lay boards were thought best able to speak for local needs. When such citizen boards were asked to define their priorities, they not surprisingly named preventive, family-oriented approaches.

To their credit, many CMHCs did reach out to the chronic patient, established relationships with state hospitals, and developed sheltered living and work arrangements for patients judged incapable of more dramatic rehabilitation. It is an irony of the program that centers would ultimately be criticized not only for ignoring the discharged patients of state hospitals in favor of patients for which funding and authorization existed, but also for creating comprehensive sheltered programs if these did not serve to return chronic patients to community life (Mollica, 1983). The clear message was that patients were supposed to get well.

Statistics now indicate that the marked decrease in the resident population of state mental hospitals between 1955 and 1975 did not reflect successful rehabilitation of patients but rather a "philosophy of short-term hospitalization" and a cycling of patients from hospital to community and back. Admission rates to state hospitals (in contrast to residency rates) remain very high, and "about half of the released inpatients are readmitted within a year of discharge" (Bassuk and Gerson, 1979: 49).

Most CMHCs neither rehabilitated the very ill nor provided round-the-clock community care. When we asked one CMCH director what his center did for the chronic population, he said, simply,

> We're not funded to care for that population. . . . Of the $4,000,000 in our budget we have the flexibility to work with the chronically ill with only a few hundred thousand dollars. [Actually] we are reluctant to care for the chronically ill. This is a very dependent population. I don't think the people who criticize the centers for not serving this group realize how much is involved. For it to be possible, the money would have to follow the patient. And it doesn't [Interview, 1978].

Insofar as the shapers of the federal act concerned themselves with patients who would remain chronically ill, they intended that *state* dollars saved through reductions at and closings of state hospitals would be transferred to CMCHs — that is, would "follow" released patients. There are a number of reasons this did not occur.

Important among them was that states, while probably avoiding the expenditure of new dollars on state hospital renovations and expansion, did not actually save many dollars. The complete closure of state hospitals, which would accomplish savings, was often contrary to the political interests of legislators from communities economically dependent on the hospitals.

It was also important that state governments tended not to embrace the CMHCs as their own. This lack of identification was often explicit. Rose (1979: 440) notes that a comprehensive study of public mental health services commissioned by the state of New York in 1976 "did not even mention CMHCs as pertinent to public mental health care." The same dissociation was evident in California. At the time of our research (1978), California did not include the federal CMHCs in its state mental health plan, but only those entities which it funded under its own Short-Doyle Act. Many officials of state departments of mental health interviewed in our research echoed the sentiment behind these planning omissions: The private-sector federal grantees in their states simply were not part of their states' plans for the mentally ill. The worst-case result was common. The director of a CMHC that consisted of a consortia of providers having a combined budget of $11 million told us, "The state hospital doesn't even tell us when it releases residents of our area." And, of course, money did not follow, either.

The general unwillingness of states to underwrite the costs of chronic care in the community has left CMHCs without predictable funding for this purpose. What federal money was available for comprehensive CMHC operations declined by formula stipulation on an annual basis. After eight to twelve years, it reached zero, and CMHCs were said to have "graduated" from federal funding. Funding alternatives for the centers have usually been contracts with county departments, industries, and, of course, categorical grants from special-purpose federal agencies. These grants — for alcoholism and drug abuse treatments, geriatrics, and so forth — have increasingly shaped the services offered by CMHCs and limited their ability effectively to serve patients who need multiple services. One geographically remote center that we visited was obviously serving chronic patients in spite of heavy dependence on categorical funds. When asked how, the director replied, "It's not all bad being [so far from the urban centers]. We're a long way from [our DHSS regional office] and a long way from Washington, and it's all uphill." This center prided itself on its accountability to patients rather than to dollars.

Categorical funding is not, of course, the only thing leading to the fragmentation of services which, for the chronic patient, must be integrated. Professionals tend to want to treat patients with the particular problems they are trained to treat, not the patient's related problems or the myriad residual problems that for some patients are central. Like their Alabama counterparts mentioned earlier, state hospital psychiatrists at the Worcester State Hospital in Massachusetts wanted to eliminate nonmedical admissions (Myerson, 1980; Morrisey and Goldman, 1980).

The psychologists in the remote CMHC just mentioned rebelled when asked to provide services they considered inconsistent with their professional training. And so on. When professionals who dominate organizations pick and choose among patients and among the problems those patients present, they are true to their professional callings and competencies as well as their categorical grants, but not necessarily to patient needs.

Professional conflicts among the staff members of single-organization CMHCs presented one level of difficulty. Problems are exacerbated when necessary dealings are with units not included within the administrative structure and when the CMHC is itself only a consortium of service providers. When services are both medical and social and are scattered among individually governed organizations, who decides and who implements becomes a major political problem. The patient's need for organizational coordination is made plain by Bassuk and Gerson (1978: 50):

> Time and time again we see patients who were released from state hospitals after months or years of custodial care; who then survived precariously on welfare payments for a few months on the fringe of the community, perhaps attending a clinic to receive medication or intermittent counseling; who voluntarily returned to a hospital or were recommitted (which in Massachusetts is possible only if the patient is acutely suicidal or homicidal or manifestly unable to care for himself); who were maintained in the hospitals on an antipsychotic medication and seemed to improve; who were released again to an isolated "community" life and who, having again become unbearably despondent, disorganized or violent, either present themselves at the emergency room or are brought to it by a police officer. Then the cycle begins anew.

Hospital, clinic, counselors, courts, the emergency room, welfare payments, police. All these must somehow work together.

But how? The CMHCs we selected for field research had reputations for excellence, and most combined an array of services under one administrative structure. Where they did not, there was trouble. In one of the communities there was so much dispute over the role of medical education on the psychiatric floor of the general hospital and over the care to be provided to indigents that the CMHC and the medical school ended up splitting the floor down the middle for purposes of administration. In another community, a major dispute surrounded whether violent patients should be restrained at the county general hospital, where they could be monitored for a variety of medical problems and complications, or at the CMHC, where they could most easily receive psychological counseling. In this case, the two institutions sat on the same hospital grounds, separated by only a small creek.

It is notable that medical science offers little clarification. Mollica (1983: 371) notes that the role of public psychiatry is obfuscated by "the profession's lack of knowledge of the genetic and neurobiological bases of psychiatric disorders" and its consequent inability to "define its legitimate treatment goals." Thus the members of each professional group, with the resources available to them, treat the patients that fit their professional criteria and funds. Remaining problems and people shuffle between agencies and personnel; coordination is usually the "patient's work" (Strauss et al., 1982).

Recent proposals for community support systems place responsibility for integration with local governments. Typically, county government, or a county-level unit, is directed by state law to sort through categorical grants and find vendors to provide needed services and oversee the result. The extent to which such units will be able to coordinate providers and competitors for control remains highly problematic. In Los Angeles County there are five county planning areas, twelve mental health regions, twenty-five mental health districts, and fifty-two federal catchment areas.

Beyond this, it appears that once again the local public general hospital is reassuming its historical role as repository of last resort — for ambulances carrying indigent ill, for families, and for police. In many of the sites we visited, local hospital authorities railed against the CMHCs and other categorically funded providers for "creaming" preferred patients and leaving the remainder on the streets or to the catch-all public hospitals.

The latest shift in policy emphasizes service contracts between local or regional bodies and private purveyors of categorical services. Such a shift of responsibility to counties and the private sector seems less of a solution than a trick for making the problems go away. Many may be responsible to specific contractual arrangements, but few will be committed to the integration of care. The chief of a medical center serving the chronically mentally ill told us,

> We can't get anybody into . . . any of our contracting hospitals if they require both medical and psychiatric care. The contract we have . . . doesn't pay for medical. . . . So these people get routinely put into the county hospitals, the only place where these bureaucratic mechanisms will not completely thwart you [Interview, 1978].

The bigger and richer the urban area, the more opportunities to fall between providers — the more invisible the sick person. Patients drift from rural areas that lack services (fewer than half of NIMH's 1500 designated catchment areas ever developed CMHCs) to urban areas that lack organization. One irate county health administrator gave his view of the resulting situation.

> What do people need? Bed and board. Medical. Recreation. . . . If you start thinking about what people need to become mentally rehabilitated, it's fairly simple. You can assign numbers to these things if you are doing it in a total institution. In a total institution, which I am not necessarily fond of, nonetheless, the cost is calculable. If you spread it around it costs more than it would cost in the total institution. But people argue that it costs less. That's because they don't figure in the amenities. The patients are supposed to take care of themselves in spite of the fact that the neighbors will come back to you and say that 'these people' are not wanted in public parks.

> The advantage of decentralization to people who don't want to pay the bill for care of the mentally ill is that you make the problem less visible. At the state hospitals, you could see the snake pit. But you did know that the patients had food, shelter, exercsie. . . . Nobody is guaranteeing that much now [Interview, 1978].

It is obvious that CMHCs have provided many needed services that previously were not available. As the president of the board of one rural center said to us, "We look at each other so often around here and say 'What did we do before Mental Health [i.e., the CMHC]?'" No doubt a certain number of the persons thus served avoid not only misery but the need for more expensive care and perhaps institutional confinements. But those being served probably are, as has been charged, a new group and, while needy, less needy than those for whom services have disappeared. What is the cost of reassuming the trusteeship of the helpless? Of offering human care either within or without an institution? One mental retardation professional put it vividly:

> I think there will be a lot more problems in the community in the future. My center, Child Evaluation, is a mental retardation unit. We are aware of many disasters: one of our children died Saturday. It was never mentioned in the paper that this child, who died in an acute care hospital, was a patient of ours. Nobody pays any attention if something like this happens in a private home, or a hospital but if it happens in a group home. . . .

> There have been two other disasters in our facilities lately: one child jumped out of a second story window and one burned in a shower with too hot water. Nobody has any idea what it would cost. . . . Nobody has any idea of what proper care would cost. . . . I'm on a state committee. My last comment upon leaving was that there wasn't enough money in the budget proposed to pay for paperwork [Interview, 1979].

THE ELDERLY: DEINSTITUTIONALIZATION CARE IN THE 1980s

It is rare today for the state mental hospital to be the focus of exposés showing abuse of the ill and helpless. Reporters and reformers have found their new material in the streets and back alleys, where, it is charged, former mental patients scavenge and die, and in the custodial institutions of our era: nursing homes. That the focus of institutional scandals would merely shift was perhaps predictable, for in a great many cases patients were merely shifted. For most of the patients discharged from state mental hospitals after 1955, deinstitutionalization was never more than "discharge to a nursing home." An observer in one county we studied described her county's institutions this way: "And next to the county hospital is the nursing home. Those patients are what we used to call mental patients." That such discharges were ever confused with deinstitutionalization shows either disingenuousness or a bureaucratic preoccupation with reductions in mental hospital censuses per se (Bassuk and Gerson, 1978). In fact, across the country large numbers of patients were not discharged at all but merely relabeled as befit new goals and new funding arrangements.

The "back wards" of the mental hospitals (patients not expected to respond to treatment) were sometimes synonymously labeled "geriatric wards" (Morrisey and Goldman, 1980: 277). As we have noted earlier, many state hospital patients were more old and confused than sick. They resided in mental institutions because the institutions took them in — whether admitted, committed, or abandoned. When the authority of the mental hospital to retain such persons narrowed, the aged and disabled were discharged. Parallel actions of the federal government, however, had opened the nursing home doors.

Nursing homes in the twentieth century, like mental asylums in the nineteenth, have depended, although less directly, on public funds intended to aid the truly needy without benefiting the remainder (Rosenblum, 1979; Trattner, 1974). The problem of making this distinction has plagued urbanized society since the industrial revolution and remains unsolved, with the result that policy for dependent populations swings back and forth between policies intended to improve conditions for the truly needy and policies (or funding decisions) designed to discourage the undeserving from applying. Aid to the elderly is no exception.

The "outdoor relief" of the British poor laws became unpopular at the time of the industrial revolution, since, so nearly equal to wages, opponents feared it would diminish work incentives. "Indoor" or institutional care was only a partial answer, even when advocates of reform associated it with illness and linked benefits to medical certification. High costs reinforced by the suspicion that the undeserving poor might somehow benefit consistently limited the level of improvement achieved for the "deserving." A case in point is President Franklin Pierce's veto in 1854 of a bill that would have provided federal land for the construction of mental asylums. Trattner (1974: 58-59) explains the veto this way:

> The promise of a continued, and perhaps expanded social welfare program stimulated by federal funds was shattered by a veto from the pen of President Franklin Pierce. While expressing "the deep sympathies in [his] . . . heart" for "the humane purposes sought to be accomplished by the bill," Pierce felt compelled to veto it because it was illegal: "If Congress has the power to make provision for the indigent insane . . . it has the same power for the indigent who are not insane," and thus all the nation's needy.

Conditions in public institutions reflected similar suspicion that the merely needy were not above preferring charity to work, if the charity compared favorably to any workaday alternative. Thus, less than adequate institutions were increasingly the objects of scandals and exposés. As a result, by the time of the Great Depression, public and professional opinion had shifted strongly against institutional care for the needy elderly.

The terrible condition of care in almshouses and asylums affected policy for the elderly in two ways: (1) It served as a powerful argument for old-age relief benefits in the form of insurance (social security and old-age assistance) and (2) when the Social Security Act was passed, a provision was included barring Old Age Assistance funds from going to any "inmate of a public institution" (Vladeck, 1980). The politically pragmatic reformers who spearheaded the drive for social security shied away from provisions for permanent disability but were resolved not to allow their hard-won program to finance the deplorable care available in public institutions. The paradoxical but predictable result of the policy was, however, new and extensive reinstitutionalization in the new form of (mostly) proprietary nursing homes:

> But though most of the infirm elderly already in almshouses remained there, restrictions on OAA payments forced others to turn elsewhere. Public facilities began to be supplanted by proprietary homes for the elderly. Social workers barred from placing the least competent and most needy of their clients in public institutions turned to proprietary ones. Moreover, many OAA recipients who had taken care of themselves or had been maintained by their families now had the purchasing power to obtain institutional care on their own [Vladeck, 1980].

Nursing homes grew slowly and steadily through the postwar period, receiving some impetus to growth through the 1950 Hill-Burton Act, which granted construction subsidies to nursing homes operated "in conjunction with a hospital." In 1960s, the Kerr-Mills Medical Assistance to the Aged bill was passed; the law was a conservative response to national health insurance for the aged proposed by Senator Kennedy. Kerr-Mills allowed states to establish "comprehensive health insurance programs for the needy aged . . . without burdening taxpayers with the expense of insurance for the nonindigent elderly" (Vladeck, 1980: 45). Although the act had limited impact on the health system as a whole, it resulted in a dramatic increase of the nursing home population resulting from a system of direct payments to nursing home operators (vendor payments) for welfare clients. These vendor payment programs under Kerr-Mills were administered by the states under broad federal guidelines. During the period 1960-1965,

> the increasing supply of nursing home beds provided welfare and hospital social worker a solution to the problems of what to do with elderly beneficiaries who had problems of any one of a number of kinds (social isolation, mental illness, even chronic disease). But even as nursing homes filled up with welfare patients, more and more were brought to the attention

of welfare agencies, sustaining the "shortage." So the vendor-payment totals grew without making a visible dent in the perceived backlog of demand for nursing homes [Vladeck, 1980: 47-48].

The architects of the next major social insurance legislation, Medicare, were aware of the difficulties a fledgling health insurance plan would have coping with the problems of long-term care. They carefully designed the proposed new health insurance for the aged to cover nursing home care only under very restricted conditions of medical convalescence. To their surprise, Medicare's longtime opponent, House Ways and Means Committee Chairman Wilbur Mills, proposed in 1965 substantial amendments to Medicare at the last minute, the most important of which was Medicaid. Medicaid was an extension of the earlier Kerr-Mills concept of means-tested medical care for welfare clients and the medically indigent. The sponsors of Medicare, who all along were advocates of a more comprehensive national health insurance than they thought politically possible, accepted Mills's support and his amendments readily, but they were not conceptually prepared for the intricacies of implementation. This was particularly true of the nursing home benefit:

> Nursing homes couldn't very well be left out of a program extending Kerr-Mills. But as no one really knew what to do about them, they were included without much definition. . . . Everyone agreed that nursing homes were inadequate, but that providing a higher level of care to those primarily in need of "custodial" services would be frightfully expensive. Yet Medicaid contained no qualitative specifications at all, while it took the budgetary lid off the vendor-payment program [Vladeck, 1980: 51].

Two main levels of care are defined under existing regulation; these, in turn, are subject to further subdivision. A skilled nursing facility (SNF) provides services under physician order that require the services of professional personnel or are performed under the supervision of professionals. An intermediate care facility (ICF) is for individuals who do not require the degree of care and treatment of a hospital or SNF, but due "to mental or physical condition require care and services (above the level of room and board) which can be made available to them only through institutional facilities." Medicare covers only SNF care, and only under fairly strict criteria restricting coverage to convalescence from hospital procedures:

> With the exception of patients with end-stage renal disease.
> . . . Medicare eligibility at present means nothing with re-
> spect to long-term care. The Medicare message to the average
> patient is clear: "Get well or get lost" [Somers, 1982: 221].

It is consequently not surprising that Medicare payments account for only 2 percent of nursing home expenditures. Medicaid, on the other hand, covers 48 percent of the nation's nursing home bills.

Medicaid allows greater coverage for long-term care, but only for the provably very poor. Medicaid, too, has provisions against the coverage of "custodial care" and requires that nursing home benefits be covered only when the services are certified by a physician to be medically needed. The nature of medical need for people whose problems are multiple and interacting can be a matter for constantly shifting bureaucratic and political decisions. Bruce Vladeck calls this the "game of levels." The "levels of care" definitions that are used in the placement of nursing home residents have to do with medical criteria, whereas the needs people have that lead to institutionalization are multiple:

> The interactive, circular nature of long-term care problems
> needs emphasis. When an individual is referred for long-term
> care, the cause of the observable functional impairment may
> not be obvious. Thus, the least restrictive or least expensive
> solution will not be intuitively apparent. For example, if an
> individual is not preparing meals, one needs to know whether
> she forgets it is meal time, is not hungry because she feels
> depressed, finds cooking for one a lonely business, is finan-
> cially strapped, is physically unable to get out and shop, or all
> of the above [Kane and Kane, 1982: 9].

If one really bears in mind the many factors other than a person's disability that may lead to nursing home placement, the common research finding that many residents are "inappropriately placed, from a medical point of view" is hardly surprising. As Steiner and Needleman (1981a: 61) report from their extensive review of the research on institutionalization, "a certain level of disability is required for entrance into a nursing home . . . [but] it is not until the effects of disability are exacerbated by lack of informal supports or low income that institutionalization is necessary." The severity of a disability may ebb and flow, but lack of family and other informal support and low income tend to persist, along with the need for long-term care. The other side of the coin, also well documented in the literature (e.g., Shanas, 1960, 1979; Shanas and Sussman, 1977), is that people with fairly high degrees of disability, who would easily qualify as SNF or ICF, live at home in the community when they have adequate income and family support.

One way to interpret these facts is to conclude that there is probably a hard core of dependent people whose social situation, including income and family status, combined with psychological and medical conditions make noninstitutional living close to impossible. If one drew this conclusion, a policy consequence would be to accept institutions as a part of the landscape and to develop realistic policies to establish quality care under the assumption that the residents of institutions have multiple needs including but going beyond medical and psychiatric care. Implied in such a policy would be the notion that a decent standard of living ought to be available for people who are only slightly disabled, but extremely indigent and isolated, as well as the notion that people who are extremely dependent for medical and psychiatric reasons ought to have specialized services available, in or out of institutions.

The policy conclusions drawn today are, however, quite different from those outlined. That institutionalization results from factors other than medical condition, including social situation, has — in the face of rapidly escalating costs of institutional care — led to the conclusion that the need for nursing home beds could be significantly reduced if we were to provide more social support for disabled people in the community. This argument is given additional force by the peculiarities of our medical-oriented gatekeeping system for long-term care. The only form in which "custodial" services can be funded by Medicaid is in a nursing home. The logic of the reimbursement policy is that Medicaid, a form of medical insurance, should not be used to fund essentially social services, unless such services are specifically indicated by medical conditions and are performed under medical auspices. Given such constraints on reimbursement, there exists a perverse incentive leading to institutionalization in cases where some additional social support could prevent it. Such considerations apply only in cases in which there already exist some informal support and resources — and, of course, additional noninstitutional services will cost money.

Several demonstration projects have been initiated over the last few years to determine the feasibility and consequences of organizing community long-term care as an alternative to nursing homes. These projects include Project ACCESS in Monroe County, New York, Project TRIAGE in Connecticut, and the Community Care Organization (CCO) in Wisconsin. The three projects mentioned all obtained permission from the federal government to use Medicaid funds for whatever services were needed to maintain persons in the community. Although the program designs varied considerably, each project involved some kind of control group for the purpose of comparing measures of cost and quality. There is a considerable body of literature evaluating these and similar projects, from which some general conclusions can be drawn (Eggert et al., 1980; Greenberg et al., 1980; Hicks et al., 1981; Applebaum et al., 1980; Stassen and Holahan, 1981) No significant or consistent differences were found in the quality of life, total costs, or rate of institutionalization between experimental and control groups. In some cases, costs were found to be higher in the experimental group, and this was usually attributed to higher utilization of services, particularly hospital services (Greenberg et al., 1980). More than one project reported that in the first year, the experimental group reported proportionately lower nursing home admissions, followed by proportionately higher rates in the second year, indicating that the availability of community alternatives may have the effect of delaying institutionalization rather than preventing it. Despite inconclusive results about outcomes, these projects demonstrated a different model of care, which provides a more integrated continuum of services than has been previously available.

Despite the results of these carefully controlled projects, many states are moving ahead and organizing community alternatives for the explicit purpose of contolling costs and encouraging deinstitutionalization. These initiatives often accompany new state policies forbidding construction of new nursing home beds, placing absolute ceilings on nursing home rates, removing Medicaid reimbursement for the lowest levels of nursing home care, and periodically reviewing the appropriateness of nursing home placements.[2] All these efforts are going on simultaneously in Wisconsin.

The centerpiece of the Wisconsin reform effort is a program known as the Community Options Program (COP). This ambitious project aims "to assure that someday the option of community care is at least deliberately considered for *all* long term care clients." The program aims at giving support for "new resources, new flexibility, and new incentives for community care, while at the same time guarding against any imprudent, excessive or unwarranted expenditure of limited public dollars" (State of Wisconsin, 1981: Section 1.1). The planning process that developed this program was initiated in the wake of steadily escalating financial pressures on Wisconsin's health system. Medicaid expenditures almost tripled from 1973 to 1978. Nursing homes account for an ever growing portion of this expense; for the fiscal year ending June 30, 1981, nursing home costs were 24 percent higher than in the previous year. In addition to the financial pressure, a number of social service professionals in and out of government and representatives of advocacy groups have been pushing for the creation of noninstitutional alternatives for long-term care.

COP is an assessment and case management program administered (according to Wisconsin tradition) by social service or mental health boards in the counties. The program is aimed at people identified as nursing home candidates and attempts to arrange and manage services that could support their living at home or in the community rather than entering an institution. The state provides funds for assessment and allows the expenditure of an average amount per case equal to the average state contribution to a Medicaid nursing home bill — around $400 a month. About the only restriction on spending is that funds cannot be used to purchase institutional care.

The program was implemented in eight counties in 1982 and will be implemented in thirteen more in 1983. In light of the findings from the demonstration community care projects, the results to date in Wisconsin are not surprising. Of the 384 clients assessed by the end of August 1982, case plans were developed for 42 percent. Although 84 percent of these persons (for whom case plans were developed) "were evaluated as having social and medical needs appropriate for Medicaid to reimburse nursing home care," one-third of those sampled in a survey indicated that they had never considered entering a nursing home. An additional 10 percent reported that, although someone else may have considered nursing home entry for them, they had not considered it as an option for themselves. Here is the state's conclusion from this finding:

> This suggests that in addition to assessing clients' medical and social needs, counties should be required to verify a client's intention to enter a nursing home before they provide COP-funded community services to clients [State of Wisconsin 1983: 208].

Contrast that recommendation with the following section of the law, which takes effect after COP has been fully implemented (participating counties have three years in which to implement the program):

> If an assessment determines that nursing home care is not appropriate for a person who is eligible for medical assistance

or . . . if state or federal funds are available to support any needed noninstitutional services for the person in the community, then medical assistance reimbursement is not available for nursing home services provided to the person. An assessment determination that nursing home care is not appropriate supersedes a physician's plan for care authorizing nursing home care if the assessment determination includes review by a physician. An aggrieved person may appeal [1981 Wisconsin Statutes, S.46.27(6) (c) (1)].

One can imagine a worst-case scenario in which those who most want to stay out of nursing homes must declare their intentions to enter one before being allowed to receive alternative services, while those who would like to enter an institution could be forbidden Medicaid reimbursement because a caseworker determined that community care was in some way feasible.

Those responsible for planning and implementing COP are aware of these problems. The quoted section of the law was inserted by the legislature during debate over their objections. The man responsible for COP implementation told us,

> We were really opposed to this because it sets up COP as a gatekeeper for the nursing home, which we don't think should be COP's function. Our goal is that alternative community care plans be developed for everyone. We think that the goal of being a gatekeeper and the goal of developing a positive alternative option for people are actually incompatible goals and that either one or the other is going to govern your activities, but both can't. Whichever one does will dominate the program [Interview, 1982].

A program such as COP, given what we know about the needs for long-term care, could make a contribution by providing needed services in a flexible manner to the chronically ill and dependent who have some existing supports. The probability is high that the clientele that would seek services and benefit from them would be in large part a new one, like those that emerged when CMHCs and Old Age Assistance made new services available. Additional economic costs would result. If COP is forced to play a containment or gatekeeping role, and is forbidden to help *except* when a client would otherwise go directly to a nursing home, its ameliorative contribution may be quite limited.

Policy discussion about the care of dependent people is frequently structured by such dichotomies as medical versus social models of care and institutional versus community care. In the nineteenth century, the terrible condition of the mentally ill in the community sparked a movement toward care in asylums supervised by physicians. Recently, both institutional care and care under medical leadership have come under criticism. The new movement toward community care holds that "medical model" institutional care for dependent people is excessively costly, inappropriate to the needs of the patients who need social rather than medical care, and that it creates new problems of functional decline resulting from institutionalization itself.

Each of these claims can be well documented. What needs to be stressed, however, is that these quite accurate claims lead directly toward specific policy recommendations or conclusions only under certain assumptions, especially the assumption that care would be cheaper, more appropriate, or freer from abuse and neglect if it were organized by a different profession or provided in a different place. Consider this example, reported in Milwaukee:

> One caller to the county complained that a woman was tied to a chair and left alone all day while her grown children went to work. When a social worker checked it out, he found that the woman was indeed tied to a chair, but loosely. Beside the chair was a plate of food, a glass of water, and a television set, so the woman would have entertainment while the caregivers were away. She had been tied to the chair so she wouldn't hurt herself. Is that abuse or benevolence?

> Social workers . . . might find an elderly man in a retirement hotel living on peanut butter, sleeping on dirty sheets, and smoking in bed, with burn holes in his clothes and bedding. But should a similar situation be termed "abuse" when a person lives the same slovenly way in the home of a relative [Davidian, 1983: 21]?

If we can get away from thinking that the location of care or its professional organization is the only relevant question, we will notice that perhaps people will have similar difficulties caring for the highly dependent in a wide variety of settings. For example, a recent survey of geriatric care in the United States (Kane et al., 1981: 27) reports that, while medical care in nursing homes is known to be substandard, evidence suggests that surveys of physician offices and hospital practices show patterns of less attentive care given the elderly than younger people, controlling for illness and intensity of illness. Variables other than disability or disease seem to determine the kind of care a person gets and where he or she gets it. Differences of social resources, personal wealth, and family support will render it highly likely that under any system of categorization or reimbursement, the socially isolated will require institutional care when they become even slightly disabled, while even extremely disabled people with supports will — as they always have — often remain at home in the care of friends and family. Public policy may have a great deal to say about the kind of care available in either setting, which will depend on the quantity and quality of resources available to support care in whatever location it takes place.

George Maddox, director of the Center for Aging and Human Development at Duke University, comments on a European-visit.

In England and Scandinavia I posed the question, "Tell me how you use information to plan." The response was inevitably, "You Americans. You remain so naive." Carl Evang . . . the grand architect of the very effective Norwegian health-care system . . . in particular felt that the key issues in allocation and reallocation were not, in fact, fine tuned evidence of performance; rather they were broad political visions that could be translated into legislation and plans for financing. . . . When [Evang was] asked about his contribution to the Norwegian health-care system, [he] responded: "*My contribution . . . was to convince the legislature that they must not authorize care for which they are unwilling to insure financing and control.*" Then he added, "*You have not*

achieved that in your country," that is, in the United States
[Maddox 1981; emphasis added].

We suspect that the involuntary response of reform-minded Americans to such advice would be to shudder, for realism about cost and toughness about control are taboo in the American political culture. To be adopted, reforms must promise to save money and humanize care while skirting the issue of controls. If the genius of American politics lies in its ability to handle reform in a piecemeal, fragmentary way, its weakness is its tendency to market gradual changes as comprehensive solutions. If it weren't for out national shortness of political memory, this tendency would drive us all crazy.

NOTES

1. For a description of this research, see Greer and Greer (1980). Data collection was funded by NIMH Grant R12 MH28448.
2. See Steiner and Needleman (1981a, 1981b) for a description of such efforts in several states.

REFERENCES

Applebaum, R., F. W. Seidl, and C. D. Austin (1980) "The Wisconsin community care organization: preliminary findings from the Milwaukee experiment." Gerontologist 20, 3: 350-355.
Bassuk, E. L. and S. Gerson (1978) "Deinstitutionalization and mental health services." *Scientific American* 232, 2: 46-53.
Becker, H. S. (1964) *The Other Side: Perspectives on Deviance.* New York: Macmillan.
Boorstin, D. (1965) *The Americans, Volume 2: The National Experience.* New York: Random House.
Bowen, A. (1979) "Some mental health premises." *Milbank Memorial Fund Quarterly* 57, 4: 553-551.
Braudel, F. (1976) Chapter 6 in *The Mediterranean in the Age of Philip II,* Vol. 1. New York: Harper & Row.
Bucher R. and A. Strauss (1961) "Professions in process." *American Journal of Sociology* 66, 4: 325-334.
Cottrell, F. (1955) *Energy and Society.* New York: McGraw-Hill.
Comptroller General of the United States (1977) *Returning the Mentally Disabled to the Community: Government Needs to Do More.* Washington, DC: U.S. Government Accounting Office.
Davidian, G. (1983) "The battered elderly." *Milwaukee Journal,* April 24.
de Beauvoir, S. (1972) *Coming of Age.* New York: Putnam.
Eggert, G. M., J. E. Bowlow, and C. W. Nichols (1980) "Gaining control of the long-term care system: first returns from the ACCESS experiment." *Gerontologist* 20, 3: 356-363.

Eisenstadt, S. N. (1956) *From Generation to Generation*. New York: Macmillan.

Erickson, E. (1963) *Childhood and Society*. New York: Norton.

Friedman, E. (1982) "The 'dumping' dilemma: the poor are always with some of us." *So. Hospitals* 56, 17: 51-56.

Glaser, W. A. (1968) "Medical care: social aspects," in *International Encyclopedia of the Social Sciences*. New York: Macmillan.

Goffman, E. (1961) *Asylums*. Garden City, NY: Doubleday.

Goode, W. J. (1973) "The protection of the inept," Ch. 5 in *Explorations in Social Theory*. New York: Oxford University Press.

Greenberg, J. N., D. Doth, A. N. Johnson, and C. Austin (1980) *A Comparative Study of Long Term Care Demonstration Projects: Lessons for Future Inquiry*. Minneapolis: University of Minnesota, Center for Health Research.

Greer, S. (1982) "Citizens' voluntary governing boards: waiting for the quorum." *Policy Science* 14, 2: 165-178.

Greer, S. and A. L. Greer (1980) "Governance by citizens' boards: the case of community mental health centers," in D. Nachmias (ed.) *The Practice of Policy Evaluation*. New York: St. Martin's.

Grob, G. N. (1980) "Institutional origins and early transformation: 1820-1968," in J. P. Morrisey et al., *The Enduring Asylum: Cycles of Institutional Reform at Worcester State Hospital*. New York: Grune & Stratton.

Gruenberg, E. M. and J. Archer (1979) "Abandonment of responsibility for the seriously mentally ill." *Milbank Memorial Fund Quarterly* 57, 4: 485-506.

Hicks, B., H. Raisz, J. Segal, and N. Doherty (1981) "The triage experiment in coordinated care for the elderly." *American Journal of Public Health* 71, 9: 991-1002.

Jones, K. (1979) "Deinstitutionalization in context." *Milbank Memorial Fund Quarterly* 57, 4: 552-569.

Kane, R. L. and R. A. Kane (1982) *Values and Long-Term Care*. Lexington, MA: D. C. Heath.

Kane, R. L., D. H. Solomon, J. C. Beck, E. B. Keeler, and R. A. Kane (1981) *Geriatrics in the United States*. Lexington, MA: D. C. Heath.

Leavitt, J. W. (1982) *The Healthiest City: Milwaukee and the Politics of Health Reform*. Princeton, NJ: Princeton University Press.

Maddox, G. (1981) "Alternative models for health care," in R. Morris (ed.) *Allocating Health Resources for the Aged and Disabled*. Lexington, MA: D. C. Heath.

McNeil, W. H. (1982) *The Pursuit of Power: Technology, Armed Force, and Society Since A.D. 1000*. Chicago: University of Chicago Press.

Mechanic, D. (1969) *Mental Health and Social Policy*. Englewood Cliffs, NJ: Prentice-Hall.

Mollica, R. F. (1983$ "From asylum to community: the threatened disintegration of public psychiatry." *New England Journal of Medicine* 308, 7: 368-373.

Morrisey, J. P. and H. H. Goldman (1980) "The paradox of institutional reform: administrative transition with functional stability," in J. P. Morrisey et al., *The Enduring Asylum: Cycles of Institutional Reform at Worcester State Hospital*. New York: Grune & Stratton.

Morrisey, J. P., H. H. Goldman, and L. V. Klerman (1980) "The enduring asylum," J. P. Morrisey et al., *The Enduring Asylum: Cycles of Institutional Reform at Worcester State Hospital*. New York: Grune & Stratton.

Myerson, D. J. (1980) "Deinstitutionalization and decentralization: 1969-1977," in J. P. Morrisey et al., *The Enduring Asylum: Cycles of Institutionalization Reform at Worcester State Hospital*. New York: Grune & Stratton.

Rose, S. M. (1979) "Deciphering deinstitutionalization: complexities in policy and program analysis." *Milbank Memorial Fund Quarterly* 57, 4: 429-460.

Rosen, G. (1963) "The hospital: historical sociology of a community institution," in E. Freidson (ed.) *The Hospital in Modern Society.* New York: Macmillan.

Rosenberg, C. (1982) "From almshouse to hospital: the shaping of Philadelphia General Hospital." *Milbank Memorial Fund Quarterly* 60, 1: 108-154.

Rosenblum, R. W. (1979) "Evolution of health care financing programs for the aged: from English Poor Law to Medicare," Ch. 1 of Ph. D. dissertation, Columbia University.

Rothman, D. J. (1971) *The Discovery of the Asylum.* Boston: Little, Brown.

Schatzman, L. and A. Strauss (1966) "A sociology of psychiatry: a perspective and some organizing foci." *Social Problems* 14 (Summer): 5-16.

Scull, A. (1977) *Decarceration, Community Treatment and the Deviant: A Radical View.* Engelwood Cliffs, NJ: Prentice-Hall.

Shanas, E. (1960) "Family responsibility and the health of older people." *Journal of Gerontology* 15: 408-411.

Shanas, E. (1979) "The family as a social support system in old age." *Gerontologist* 19, 2: 169-174.

Shanas, E. and M. B. Sussman [eds.] (1977) *Family, Bureaucracy and the Elderly.* Durham, NC: Duke University Press.

Somers, A. R. (1982) "Long term care for the elderly and disabled: a new health priority." *New England Journal of Medicine* 307, 4: 221-226.

Stassen, M. and J. Holahan (1981) *Long-Term Care Demonstration Projects: A Review of Recent Evaluations.* Washington, DC: Urban Institute.

State of Wisconsin (1981) *The Community Options Program: Guidelines and Procedures.* Madison: Wisconsin Department of Health and Social Services.

State of Wisconsin (1983) *The Community Options Program: An Evaluation of Early Implementation Experience.* Madison: Wisconsin Department of Health and Social Services.

Steiner, P. A. and J. Needleman (1981a) *Cost Containment in Long-Term Care: Options and Issues in State Program Design.* Washington, DC: National Center for Health Services Research.

Steiner, P. A. and J. Needleman (1981b) *Expanding Long-Term Care Efforts: Options and Issues in State Program Design.* Washington, DC: National Center for Health Services Research.

Stevens, R. (1971) *American Medicine and the Public Interest.* New Haven, CT: Yale University Press.

Strauss, A. L., S. Fagerhaugh, B. Suczek, and C. Wiener (1982) "The work of hospitalized patients." *Social Science and Medicine* 16: 977-986.

Trattner, W. i. (1974) *From Poor Law to Welfare State.* New York: Macmillan.

U.S. Congress, Joint Commission on Mental Illness and Health (1961) *Action for Mental Health.* New York: Basic Books.

Vladeck, B. C. (1980) *Unloving Care: The Nursing Home Tragedy.* New York: Basic Books.

Webb, W. P. (1931) *The Great Plains.* Lincoln: University of Nebraska Press.

Wiener, C., S. Fagenhough, A. Strauss, and B. Suczek (1982) "What price chronic illness?" *Society* (January/February): 21-30.

6

The City and Disability

CLYDE J. BEHNEY
ANNE KESSELMAN BURNS
H. DAVID BANTA

□ VIEWING DISABILITIES IN THE CONTEXT of an environment that may be handicapping and viewing cities as environments of numerous interacting functions allows us to examine disabilities and the city in a somewhat new light.

This chapter first discusses the concepts of impairment, disability, and handicap, placing special emphasis on the role of an individual's social and physical environments, followed by a discussion of disability-related technology. Following that, we present data on demographic information on disabilities in urban versus rural situations.

The main sections of the chapter, then, cover the city as cause of or deterrent to impairment, the city and treatment or rehabilitation, and the city as a handicap reducing or causing environment.

CONCEPTS AND DEFINITIONS

IMPAIRMENTS, DISABILITIES, HANDICAPS

What is an impairment, a disability, or a handicap? Society implicitly defines a population of people with "typical" functional ability. In contrast, society defines those who cannot perform, because of physical or mental impairments, one or more life functions within the broad range of typical as "disabled" or "handicapped."

Obviously, there are many possible definitions of the terms "handicap" and "disability." Definitions are important: They affect the methods for identifying, and actual identification of, people in need of assistance. It is probably most accurate to use the phrase "having a disability" in describing a person with some type of function limitation, given no specific background information. A "handicap," in contrast, must be related to environmental and personal contexts.

Disabilities and handicaps arise from impairments, which are the physiological, anatomical, or mental losses or "abnormalities" resulting from accidents, diseases, or congenital conditions. Generally, an impairment results in a disability when a generic or basic human function such as eating, speaking, or walking is limited. It results in a handicap when the limitation is defined in a socially, environmentally, or personally specified context, such as the absence of accessible transportation to take the disabled people to work.

Because it is the most objectively diagnosable condition of the three, and because it is based on a physical or mental loss or deficiency, an impairment is the condition for which health care is the most crucial and appropriate. Medical or surgical intervention is often the earliest form of intervention applied in order to eliminate or reduce the impairment, to keep it from becoming a disability or to keep the disability to a minimum. Furthermore, the physician is usually the point of entry into the complex of services related to disabilities.

As long as the impairment exists and is not fully compensated for, however, a disability will also exist. With disabilities, the role of health care may still be important, but it will normally be supplemented by other interventions. Some of these will be quite closely related to medical care, such as training in the use of braces. Examples of other types of services that become important include attendant care, special-education services, modified automobiles, accessible transportation and housing, community services, environmental control systems, assistive communications devices, and modifications to job sites and tasks.

When limitations cannot be eliminated, a disability remains, but it may be prevented from handicapping the individual. A disability may change the way one accomplishes a task or reduce one's ability to do it at a certain level, but a handicap may prevent the person from doing the task at all or at an acceptable level. For example, when a wheelchair fits through a doorway and into an elevator, the disabled person has access and is not handicapped in relation to that functional ability. An hour later, in the next building, a doorway may create a handicap. A more subtle example is provided when a person uses a communication device (that produces artificial speech), thus speaking in a different way from nondisabled people. That person may not be able to speak as quickly or as expressively as is "typical," and thus may become handicapped by that disability in combination with social expectations for conversational style and rate.

Clearly, the environment is an important determinant of handicaps, and the fact of living in an urban setting is one of the major factors shaping an individual's environment and creating or preventing handicaps.

TECHNOLOGY AND DISABILITY

We define technology broadly, as scientific or other organized knowledge applied to practical purposes. This definition encompasses not just physical items, such as subway elevators or artificial kidneys, but also process technologies, such as vocational rehabilitation systems and the Medicare program. Technology for disabled people plays the role of improving the fit between individuals and their environments.

The influence of technology is felt in nearly every dimension of the lives of disabled people and in policies relating to disabilities. In some cases, technology is the cause of impairments. Disease due to chemical waste dumping, death and illness due to adverse drug reactions, and automobile injuries illustrate this. In other instances, technology can eliminate or reduce impairments and keep them from becoming disabilities. Knee implants and prescription eyeglasses are examples of medical technologies designed to do this. Furthermore, technology is used to facilitate "mainstreaming" in education, to prepare disabled people for employment or reemployment, to adapt the tasks and physical sites of jobs to the capabilities of disabled persons, and to create a controllable home environment. It is also used extensively to prevent disabilities from becoming handicaps by, for example, making transportation systems and accommodations accessible. Technology enters the lives of disabled people in ways that people without disabilities may consider mundane — in the form, for example, of special utensil attachments or uniformity of traffic light bulb placements.

The state of technological capability in part determines what policies (e.g., what legislation and regulations) are possible. It very clearly affects their implementation. Federal, state, and local governments have created dozens of programs that relate to the "needs" of disabled persons. There are programs (and agencies) concerned with research, income maintenance, health care, education, transportation, housing, independent living — the list continues. Being aware of the goals and operations of these programs is important because not only are they affected by the state of technology, they in turn very much affect the development and application of technologies.

The U.S. Congress's Office of Technology Assessment (OTA) recently completed a study of technology and disability (1982). That study found that there were many specific as well as systemwide shortcomings in the extent of and ways that technologies are being applied to limitations caused by disability. We draw on our experience with that study in subsequent discussions in this chapter. The major finding of the study, however, was that technology's potential for disabled people was not being met, less because of technological problems than because of social factors — inadequate and inconsistent financing, conflicting and ill-defined goals, hesitancies over the demands of distributive justice, and isolated and uncoordinated programs.

URBAN DEMOGRAPHICS ON DISABILITIES

BACKGROUND

Estimates of the number of people with disabilities are plagued by practical as well as conceptual problems. Methodological weaknesses contribute to the poor state of information, although increased attention and funds for the collection of data relevant to decision makers could be of tremendous help. Also, there is double counting of some people with more than one disability, underreporting of some disabilities (in part due to the stigma attached to being included on a list of disabled people), overcounting by organizations seeking to make a strong case for the extent of a particular disability, and incomplete counting of some disabled people, particularly those in institutions.

A perhaps more important problem with reported counts is that they usually do not take into account the severity of the functional impairment reported. Attempts to identify populations needing services should be designed to take into account severity and functional status as well as type of disability and handicap.

NUMBERS

In part because impairments and disabilities are not as objectively measurable as is desirable and because handicaps may change depending on their context, there is no dependable count of the total number of disabled or handicapped persons (U.S. Congress, Office of Technology Assessment, 1982). Despite these difficulties, and despite the fact that for many (if not most) policy-related purposes a count of the total number is unnecessary, considerable time is spent by researchers and various groups in making such estimates. Some of these estimates range as high as 45 million, including more than 10 million children. Typical lower-range estimates are from 15 to 25 million people. Higher numbers may reflect an attempt to count people with impairments. Lower ones may be reflecting attempts to count people with more significant (more severe) disabilities or handicaps. For example, one study has estimated that approximately 12 percent of children have impairments but that only about 3.9 percent have a limitation of activity (Butler et al., 1981).

The preceding problems with demographics in this area having been noted, some data on the numbers and impairments of disabled people are presented below. These data are provided primarily as examples to place disability-related problems, opportunities, and public policy issues into some perspective.

Estimates developed from the 1977 Health Interview Survey (U.S. Department of Health, Education and Welfare, 1977) provided the following numbers on persons with selected impairments: 11,415,000 blind and (at least moderately) visually impaired people; 16,219,000 deaf and hearing-impaired people; 1,995,000 speech-impaired people; 1,532,000 people affected by paralysis; 2,500,000 people with upper-extremity impairments (not including paralysis); 7,147,000 people with lower-extremity impairments (not including paralysis); and 358,000 people with the absence of major extremities. The total of the above is 41,166,000, and *there are definitely overlaps*. Overall, 67 percent of the impairments are found in the categories of blind and visually impaired and deaf and hearing impaired. Except among those over 65, there are slightly more impaired males (52 percent) than females (48 percent).

An examination of the working-age population is useful, because an inability to work because of disability often results, in our society, in income subsidization or in technological assistance to allow employment. According to the U.S. Department of Health and Human Services (1980), in 1978, of 127.1 million noninstitutionalized working-age Americans, 17 percent, or 21 million, were limited in their ability to work due to a chronic health condition or impairment. The most common work-limiting problems are heart conditions, spine or back conditions, and arthritis. While similar proportions of men and women reported some degree of disability, a greater proportion of women were characterized as severely disabled. Interestingly, the preliminary results of the 1980 census indicate a noninstitutionalized working-age population of 144.5 million, but only 12.4 million people with a work disability. More than half of these people, 6.4 million, were prevented from working at all due to their disabilities (American Demographics, 1982).

The prevalence of disability increases with age: Adults between the ages of 55 and 64 were ten times more likely to be severely disabled than adults aged 18 to 34 (U.S. Department of Health and Human Services, 1980). Severe disability was almost twice as prevalent among the black population as among members of other races. This higher prevalence among the black population is apparent only when all disabilities are considered. If a particular condition is viewed separately — cerebral palsy, for example — the prevalence may be higher among the white population (Sigelman et al., 1977).

Self-reported "transportation disabilities" also range in the millions. Preliminary data from the 1980 census indicate that more than 6.2 million people have mobility problems that limit or prevent them from using public transportation.

Generally, disabled people are poorer and less educated than the nondisabled, and this is particularly true in the case of those who are severely disabled. However, statistics on average earnings and levels of education can be deceptive, based as they often are on people in or known to public- and private-sector programs. Thus, those people who are less educated and who earn less are those most likely to be counted. This does not mean that there is no problem of low disposable income or of educational level among disabled people. It merely implies that the most successful disabled people may be counted less, with implications not only for resource allocation and statistical bases but also for the development and maintenance of stereotypes and attendant attitudes.

URBAN DISABILITY NUMBERS

Given the poor state of disability data in general, it is not surprising that there is a shortage of helpful data on disability prevalence broken down by metropolitan versus nonmetropolitan or similar categories. Comparability of the data that do exist is shaky, to say the least.

It appears reasonably safe to say, however, that the porportion of disabled people living in standard metropolitan statistical areas (SMSAs) is roughly equal to the proportion of the general population living in SMSAs (U.S. Department of Health, Education and Welfare, 1979a; Donabedran et al., 1980). Approximately 68 percent of the U.S. population was living in SMSAs in 1977. Of the 41.2 million disabled persons mentioned above (forgetting the overlaps, for the purposes of calculating ratios), about 26.5 million, or about 64 percent, lived in SMSAs. Given the quality of the data, we do not believe it is appropriate to read too much into the slightly lower percentage in SMSAs, particularly since many people believe that disability in highly urbanized populations is underreported to a greater extent than in other populations (U.S. Department of Health, Education and Welfare, 1976).

Except for absence of major extremities, where the proportion in non-SMSAs was substantially greater than one-third, most specific categories of disability follow the same proportion. Table 6.1 provides estimates of numbers of individuals with selected impairments and their percentage distribution by

TABLE 6.1   Estimates of Individuals with Selected Impairments,
             United States, 1977, and by Percentage in SMSA
             and non-SMSA

Numbers (in 1000s) and Residence	Blind and Visually Impaired	Deaf and Hearing Impaired	Paralysis	Orthopedic Impairment	Absence of Major Extremities
Estimate	11,415	16,219	1,533	9,648	358
% SMSA	64.2	62.6	61.7	67.8	57.5
% non-SMSA	35.8	37.4	38.3	32.2	42.5

SOURCE: Adapted from U.S. Department of Health, Education and Welfare (1979a).

SMSA versus non-SMSA. About 16 percent of people aged
18-64 not living in SMSAs have disabilities that limit their
ability to work or prevent them from working. For the SMSA
population, the figure is 12 percent (U.S. Department of
Health, Education and Welfare, 1979a). The prevalence of
specific disabilities giving rise to the work limitation follows the
same pattern of being lower in urban areas. This holds true in
general for mental disabilities as well as physical ones, with the
interesting exception of severe emotional disturbances, where
the proportion is higher in urban areas.

The U.S. Congressional Budget Office estimated in 1979
that 13.4 million Americans were handicapped in the use of
public transportation because of either disability or age and
that 7.4 million of these people lived in cities served by public
transportation, or 55 percent of the total. This is a difference of
approximately 9 percent from the 64 percent of disabled people
living in SMSAs. This significant difference probably cannot be
accounted for by differences in definitions. For example, if
"living in cities served by public transportation" refers to more
areas than just those in SMSAs, the figure would be even lower
than 55 percent if it were adjusted to include only SMSA
residents. Our impression is that the figure may represent the
greater accessibility to usable transportation in SMSAs, particularly in non-inner-city areas. Thus a disability is less likely
to result in a transportation disability.

In summary, disabilities appear to be divided roughly in proportion to population between urban and nonurban, with slightly less prevalence in urban areas, especially for transportation-related handicaps and work-limiting disabilities. Although there are few data on which to base strong statements, our interpretation of the situation is that impairments do not differ significantly between urban and nonurban, but that disabilities are often less handicapping in urban settings.

THE CITY AND PREVENTION
OF IMPAIRMENTS

PREVENTION OF IMPAIRMENTS

Most people would agree that prevention is the theoretical ideal. However, despite many significant success stories (such as polio vaccine), the potential of prevention remains unfulfilled due to a combination of inadequate knowledge, human nature, and finite resources.

Even if resources were unconstrained, it would be difficult to prevent diseases and other disabling conditions whose causes are unknown or for which no effective preventive technologies can be devised with current knowledge. Furthermore, different kinds of knowledge are required to deal with the various causes of disability. For example, knowledge about the prevention of accidents is only one form of information needed, and it is quite different from the knowledge needed about preventing diseases. This knowledge constraint underscores one of the aspects of "resource capability": Resource capability refers not just to the *amount* of resources available but also to the degree of *ability* to use them.

Research, especially basic biomedical research, is creating promising possibilities for prevention. The areas of motor function, tissue structure and regeneration, molecular genetics, enzyme function, and cell biology in general are examples promising dividends in terms of disease prevention. Just as important, however, is research on the development and engineering of technologies whose science base is already established. This type of research and the resulting technologies are, in fact, the important ones in terms of prevention and the city. Research is needed in services delivery (e.g., to determine how vaccines can be most effectively delivered, especially to those in the inner city), in policy and programs (e.g., to determine the respective roles of the city, county, and state in funding and administering prevention programs), and in demographics and epidemiology (e.g., to determine who is at risk and how such individuals can be identified in advance).

In discussions of prevention, human nature is sometimes termed "imperfect" or "self-defeating" because humans do not always seem to act in their own, safe, rational best interest. Although the philosophical dimensions of the attitudes behind the use of such terms are not the subject here, it should be noted that seemingly irrational, risk-taking behavior is not necessarily imperfect or improper. It may be a reflection of different individuals placing different values and interpretations on risk-taking, risk aversion, the probability of negative outcomes, and the meaning of possible outcomes. What should be of interest to urban planners and administrators is whether residents of cities have values in relation to risk aversion or prevention that are different from those of people in general.

Policies toward prevention must take human nature and the characteristics of the populations involved into account. The success of a city's public health campaign of immunization against childhood diseases, for example, is very much dependent on the willingness and ability of parents and children to comply, to take the vaccine. This may require making strong efforts to inform residents of the existence of programs and of the risks and benefits of vaccination. It may also require the establishment of various sanctions for failure to comply. For example, all states have laws that do not allow children to begin a school year if they are not properly immunized, and most states have extended these laws to licensed day-care centers and the upper grades of secondary school.

Preventive technologies are applied by all individuals and all institutions in society. Individuals apply them when, for example, they seek prenatal care, wear seat belts, stop (or never start) smoking, reduce their use of drugs, and follow safety instructions on the job. Institutions apply them at all levels. The federal government devotes resources to auto safety, maternal and child health, immunization campaigns, food and drug safety regulation, and airport safety. States and local governments apply resources to similar activities, including health and safety regulation in the workplace and in public institutions such as schools. Industry and other commercial organizations can apply preventive technologies in workplace safety and in programs for alcoholism and drug abuse. Schools teach driver education. The list could, of course, go on.

Because resources are finite and often quite limited, they must be distributed between current treatment and rehabilitation of disabilities and the prevention of them. This is especially true today for local governments with tightly constrained budgets.

Thus, even if analysis of cost and benefits of prevention versus treatment and rehabilitation indicates that prevention is economically and humanistically preferable, the current existence of people with disabilities means that not all resources can be allocated to prevention.

PREVENTION AND THE CITY

Children ages 1-4 in inner cities are the least likely group to be vaccinated against disabling conditions such as polio or rubella. In 1979, only 58 percent of these children in the inner-city portions of SMSAs were vaccinated against rubella, and only 52.1 percent against polio (U.S. Department of Health and

Human Services, 1981). In poverty areas of inner cities the percentages were only 52.8 and 44.5. Non-inner-city and non-SMSA children ages 1-4 were about equal in their vaccination rates, between 60 and 70 percent, depending on the type of vaccination. Although we understand that vaccination rates for ages 1-4 may have increased significantly since 1979, the disparity between groups has almost certainly remained.

This discrepancy holds true for many preventive technologies in addition to vaccination, such as prenatal and well baby care (U.S. Congress, Office of Technology Assessment, forthcoming). The use of preventive services by children in low-income urban areas seems to depend on access to a regular source of care (U.S. Department of Health and Human Services, 1981). Neighborhood health centers and public health clinics may be effective alternatives to private physician offices. In contrast, one problem is that many lower-income urban residents use the outpatient departments of hospitals as their source of care. For example, residents of SMSAs used hospital outpatient departments for physician visits at a 40 percent higher rate than did residents of non-SMSAs (U.S. Department of Health and Human Services, 1981).

Opposing these figures, however, is a large body of evidence that health status is generally poorer in nonmetropolitan areas. Accidents, for example, occur far more frequently in non-SMSA areas. For 1968-72, mortality from motor vehicle accidents was 70 percent higher in rural areas than in urban ones (U.S. Department of Health and Human Services, 1981), resulting in part from safer (i.e., more preventive) traffic usage and facilities and in part from greater availability of emergency medical services. More information on the availability of health services to keep impairments or potential impairments from becoming disabilities is presented later in this chapter.

Apparently, residents of non-inner-city areas of SMSAs use medical services-related preventive services and regular primary health care more than do residents of non-SMSAs or residents of inner cities. The gap in usage rates between the "suburban" population and the inner-city population is sufficient so that, taken as a whole, the SMSA population use rate is higher than that of the non-SMSA. When this greater use of medical and health care-related preventive services is added to the other impairment-preventing aspects of urban life, the impairment-causing aspects may be (somewhat) more than compensated for — in the aggregate.

Some of the aspects of urban areas that reduce the incidence of impairments have been noted: vehicle accident-reducing techniques (such as better traffic control, greater use of mass transportation systems, lower average speeds), greater numbers of emergency medical technologies (ambulances, hospital emergency departments, paramedics), and more frequent regular visits to a physician. Another aspect may be the higher concentration of specialized medical treatment facilities and trained medical specialists. Similarly, areas of concentrated population can take advantage of economies of scale and availability of expertise for things such as enforcement of workplace health and safety regulations, alcoholism counseling programs, community health (and mental health) programs and clinics, disease screening programs, hotline counseling services, and community nutrition and food services.

Working in opposition to these factors, however, are elements of urban life that directly or indirectly increase the incidence of impairments: concentrations of poverty and poor nutrition, stress, crime, drug use, sometimes excessive noise levels, air and water pollution, and greater pedestrian traffic.

In addition, disability prevalence rates of urban areas are increased because of concentrations of subpopulations with genetic predispositions to sometimes disabling conditions (such as Tay-Sachs disease among Jewish people or Cooley's anemia among those of Mediterranean descent).

Urban governments should spend time seriously examining impairment-causing or -preventing aspects of life in their areas.

Considering the long-range costs involved in making cities accessible and supportive for disabled people, decisions regarding short-term expenses for prevention are critical.

THE CITY AND TREATMENT
OR REHABILITATION OF
IMPAIRMENTS AND DISABILITIES

Treatment and Rehabilitation. Treatment and rehabilitation of impairments so that they result in a minimum of disability depend (as does prevention) on the amount of resources available and the ability to use those resources. For example, a person with a mild hearing impairment might not have a disability if an appropriate hearing aid were available for purchase, if funding permitted that purchase, and if it effectively reduced the amount of hearing loss. Without any one of these conditions, the individual would probably be disabled.

The same principle holds true for the treatment or rehabilitation of disabilities. Given a disability, it is desirable to provide treatment or rehabilitation so that in its interaction with the environment the disability does not become a handicap. Treatment and rehabilitation resources must be available, *and* they must be appropriately utilized by those who need them.

Amount of Treatment/Rehabilitation Resources. In keeping with the generally inadequate level of data in the disability area, information on the amount of treatment or rehabilitation resources by location of residence is scarce and of less than optimal quality. Nevertheless, there is some available information that helps to provide a picture of the amount of available resources.

The types of resources needed to treat or rehabilitate impairments and disabilities naturally vary according to the specific impairment or disability, but they may be generally categorized as follows: medical treatment; physical and mental restoration services; vocational training and rehabilitation; physical rehabilitation; special education; rehabilitative devices, including prosthetics, orthotics, mobility aids, and telecommunications and other sensory aids; interpreter services for deaf people; readers for blind people; and counseling and guidance services.

Although medical treatment is not necessarily the most appropriate type of service to prevent an impairment from becoming a disability or a disability from becoming a handicap, it is often the type first received by individuals needing assistance. Because of the nature of our medical care delivery and insurance systems, physicians usually provide or prescribe rehabilitative technologies, in addition to standard medical treatment.

Medical treatment of people with impairments or disabilities may be provided by several types of physician specialists. The rehabilitation physician (physiatrist) however, is trained to perform the broadest range of rehabilitation services (President's Committee on Employment of the Handicapped, 1978). Currently, there is a national shortage of physiatrists, and this shortage is predicted to continue. For example, estimates for demand in 1990 range from 4,000 to 4,900, while estimates for supply range from 3,380 to 2,900 (President's Committee on Employment of the Handicapped, 1978). It is likely that this shortage is less severe in the cities than in other areas — the total number of specialists per 100,000 persons increases with increasing urbanization (Donabedian et al., 1980).

In general, people in urban areas have better access to physicians, and as noted previously they take advantage of that access more frequently. Between 1976 and 1978, individuals in metropolitan areas had an average of 5.0 physician visits per year, while those in nonmetropolitan areas had 4.5 (U.S. Department of Health and Human Services, 1981). These data may indicate a greater availability of physicians to provide treatment for impairments or disabilities in urban as opposed to nonurban areas.

It appears that, in addition to having access to more medical treatment, disabled people residing in cities also have more of other types of resources. Reports of the unavailability in rural areas of social services relevant to rehabilitating impairments or disabilities are an example of the evidence for this conclusion (Hulek, 1969).

Generally, the amount of resources in the city is more conducive to treatment or rehabilitation than that in nonurban areas. Urban areas have a higher proportion of academic medical centers, major hospitals, and, most important, comprehensive rehabilitation centers. However, mere physical availability of these resources guarantees neither the ability to use them nor their actual use.

ABILITY TO USE
TREATMENT/REHABILITATION RESOURCES

Several factors affect the ability of an individual with an impairment or disability to take full advantage of treatment or rehabilitation facilities or services: availability of funding, coordination of services, access to information on available services, gaps in enrollment for public or nonpublic programs, and the difficulty in maintaining medical/rehabilitative devices. As will be discussed further, the urban environment often has a decided influence — either positive or negative — on those factors.

Perhaps the most important factor determining a person's use of treatment or rehabilitation services is the availability of funds to pay for those services. Many disabled people are poor. Indeed, three-fifths of disabled adults of working age earn incomes at or near the poverty level (Bowe, 1980b). Therefore, for most disabled people, the main sources of funding are the public programs for which they are eligible. The adequacy of funding in these public programs for treating impairments or disabilities, the consistency of funding/programs among states or regions, and the means by which eligibility is determined are important issues worthy of discussion in another context. The important point here, though, is that eligibility for public funding does not depend to any significant extent on location of residence; thus, funding may be adequate or inadequate generally equally for metropolitan and nonmetropolitan residents. However, it is unclear whether the varying dollar levels of program benefits of state and local governments offset the varying local costs of living. There may be an urban/nonurban differential in some areas. For those who receive private assistance, it is likely that the same principle generally holds.

Another factor that affects the use of existing resources is the extent to which treatment or rehabilitation resources are coordinated. A common problem, often raised in the literature and in personal interviews, is that services for disabled people come from so many different, often uncoordinated, sources that users and providers are either unable to take advantage of available technologies or must spend enormous amounts of time providing the coordination needed to assist each individual to best advantage. This problem is particularly evident in the city, where the choice of services is greatest. For example, a survey of health care services for people with arthritis in an urban area documented striking underutilization and lack of coordination and communication among existing community services (Liang et al., 1981). Urban policymakers need to pay far more attention to improving coordination of services than most of their rural counterparts, who have fewer services and a smaller number of programs in their communities.

Counterbalancing (to some uncertain or unknowable degree) this difficulty in coordination is the possible higher quality, on average, of the health care services received in urban areas, due in part to the greater availability of academic health centers and other high-quality institutions, to the greater concentration and availability of highly trained general and specialty practitioners, and to the benefits that arise from access to a greater range of available services. Thus, what is in some respects a problem may also be of substantial benefit.

It is only logical that, for an individual to receive treatment for an impairment or disability, he or she must be aware of the available resources. Access to such information, however, is often a problem, particularly if the person is not eligible for public or private programs, for it is through those programs that most of the information on available resources is transmitted (U.S. Congress, Office of Technology Assessment, 1982).

Disseminating information about rehabilitation and other services is difficult no matter what the setting. Whether urban areas have an advantage or a disadvantage in this regard depends on the balance between at least two factors: On the one hand, because of the density of people in cities (resulting in a greater opportunity for some of them to be "missed" in any outreach efforts) and the sometimes confusing array of services available, city dwellers with impairments are less likely to receive information. The opportunity for preventing an impairment from becoming a disability is thus diminished for those with less access to information. On the other hand, these problems are offset (to an unknown degree) by the advantages of urban communications. Print and broadcast media, volunteer groups, and community service centers can all contribute to the spreading of information, and these are usually found in greater numbers in cities.

Given that access to information is better for those eligible for public or nonpublic programs, there are still gaps in access for those unenrolled but eligible people. Reasons for the existence of unenrolled but eligible individuals have been reported as partly due to a lack of public awareness, partly due to a lack of outreach efforts to correct it, partly due to the lack of systematic methods to correct it among uncoordinated programs, and partly due to the system's inability to handle all eligible clients due to a shortage of funds or personnel (National Arthritis Advisory Board, 1979; Stedman, 1977). Although it is difficult to identify eligible but unenrolled people whether they are in nonmetropolitan or metropolitan areas, that fact does not diminish the need for specific attention to the problem by those focusing on disability in the city.

Even for individuals who have received a medical or rehabilitative device that can prevent an impairment from becoming a disability, maintenance of that device can be a serious problem. Users must be able to obtain parts for their devices, contrive a way to function while the device is being repaired, and then pay for the whole process. For example, most users of battery-driven wheelchairs must maintain a second chair for the times when their primary chair is being repaired, because even the simplest repairs can take months. Not surprisingly, the difficulty or ease of obtaining maintenance services varies from device to device. However, residents of cities are usually at an advantage over those in rural areas. Cities are more likely to house the specialized repair centers established and staffed by the larger manufacturers. They are also more likely to have general device repair personnel available to improvise in the numerous cases of customized devices.

Taken together, the small amount of evidence seems to indicate that cities are more likely to have rehabilitation and related services and that these services are more likely to be used. However, the evidence does not permit any conclusions concerning the relative quality of services. Furthermore, it is possible that availability of services and the level of information dissemination are lower in inner cities than in other parts of metropolitan areas. As is often the case, data on the more positive aspects of non-inner-city areas may be hiding conditions in inner cities when urban areas are examined in the aggregate.

COPING WITH DISABILITY: THE CITY AS A HANDICAPPING OR HANDICAP-REDUCING ENVIRONMENT

A HANDICAP AND ITS ENVIRONMENT

An impairment leads to a disability when medical, rehabilitative, or other interventions do not provide or allow reasonable compensation for the loss of function due to the impairment. For example, loss of a limb would not result in a disability if an artificial limb were fitted that fully compensates for the anatomic loss. Even when disability results, however, a handicap is not automatically present. As stated earlier, a handicap is the result of the particular disability in combination with the individual's physical and social environments. A person in a wheelchair, for example is handicapped in terms of recreation if parks, concert halls, and movie theaters are not adequately accessible. If they are accessible, a handicap may not exist. The situation is complicated by the necessity of considering the social environment as well. Even if the physical sites of recreational facilities are accessible and do not create a handicap, attitudes and policies of the managers or employees of facilities may do so. Other users of the facilities may make it extremely uncomfortable for disabled persons to enjoy the activities, or the disabled persons themselves may feel uncomfortable in using facilities, and thus a socially derived handicap may exist. The result is the same, but the handicaps formed by physical barriers and official policies are the ones most relevant to urban governments.

This section presents some thoughts on features of urban life, organized by a few selected aspects of disabled life, that impose or reduce handicapping conditions for people with disabilities.

MOBILITY: ACCESS AND TRANSPORTATION

This is probably the area of the city life most studied in relation to disability. It is also the one most obviously oriented toward high-technology. A great many articles have attempted to calculate the relative advantages and costs of alternate means of transporting disabled people. Most have added little clarity to the debate. One common shortcoming of policy analysis in this area has been the focus on the direct function of "transportation." Too little analysis has considered transportation in the context of mobility. Mobility involves not only access to transportation but also access to destinations. It is a more systems-oriented approach than most have taken.

Cities have a natural advantage in providing accessible mobility services to disabled people: Urban areas are generally denser than rural or nonurban ones. Cities have more options open to them: They can choose some blend of mass transit, paratransit, or dedicated services to provide the transportation portion of mobility. Rural areas, for example, generally are limited to services such as demand-response or paratransit systems. Cities have a much higher proportion of buildings with elevators, many of which are already accessible to wheelchairs and all of which can be inexpensively modified to include, for example, braille markings on the elevator control panels.

Furthermore, cities can achieve a greater impact per investment from modifications of transit systems, building accessibility, and other disability-related changes because of population density. Although the proportion of people with disabilities is roughly equal between urban and rural areas, there is, logically enough, a greater concentration of disabled people in urban areas. Each public building, park, theater, and the like may be used by a higher absolute number of disabled people.

The positive contribution of city living to mobility is evidenced by the lower percentage of city residents indicating mobility disabilities (see preceding sections).

None of this suggests that there are not grave economic implications in cities making transportation systems and destinations more accessible. We believe that there is a very real possibility for future savings from reduced income maintenance, attendant care, health benefits, and so on due to increased opportunities for employment and for independent living. Although these are difficult to quantify and predict, they are still important elements of any public investment equation.

Obviously, there are many aspects of urban mobility that increase handicaps. Automobile use is more difficult for disabled people in the city (LaRocca and Turem, 1980). Disabled people who use specially modified or simply larger-size autos and vans have to contend with crowded streets, lower gas mileage (due to the size of their vehicles and inefficient rates of speed) leading to perhaps higher per mile operating costs, and difficulties in finding adequate parking spaces. These logic-based perceptions are borne out by at least one study that found less ownership and use of private cars and vans in cities than in nonurban areas (Buchanan and Chamberlain, 1978).

Another urban policy-related point is that many urban-area transportation systems, public buildings, parks, and so forth are in urgent need of repair or replacement. Moreover, as is well known, it is much less expensive to incorporate accessibility when one is already planning to rebuild or extensively modify. Any such renovation program should not miss the opportunity to make low-cost improvements in the ability of urban disabled resident to get out of their homes; to be transported to work sites, public buildings, shopping and entertainment centers, health and social services centers, or longer-range transit points (such as airports); and to enter and successfully take advantage of those locations.

EMPLOYMENT

According to the President's Commission for a National Agenda for the Eighties (1980), the rapid growth of the "service economy" is centered in urban areas. The traditional prominence of manufacturing and heavy industry in our nation's economy is being lessened. In its place are service industries such as communications, financial services, computer services, and knowledge development organizations. These types of services are often ones that can employ physically disabled people more easily than can heavy industry. Thus the city may possess an advantage in terms of employment of disabled people.

In fact, the city has traditionally been the place where nonurban individuals come for better employment prospects. This has been true for disabled people too, even when employment training/vocational rehabilitation has taken place in a rural area (MacGuffie et al., 1969). Countering this is the fact that disabled persons' geographic mobility is sometimes not as high as that of nondisabled persons.

Employment prospects are also somewhat better in the city due to the lower levels of transportation handicaps. This is not to say that there are not serious problems in the transportation/employment process; rather, it is a reflection of the higher *percentage* of disabled people who have access to transportation to the workplace.

Also, the fact that vocational training centers, jobs, sources of technological aids, and support services (such as independent living centers — see below) are often located in close proximity to one another is a potential advantage for urban areas.

Of course, the current state of the economy, with its high unemployment and low levels of capital availability, makes employment opportunities for all people more critical and also may reduce the ability of urban and nonurban governments alike to devote more resources to increasing employment of disabled people.

RECREATION AND CULTURAL OPPORTUNITIES

We found no data pertaining to urban versus nonurban areas in the handicapping or handicap-reducing aspects of recreational or cultural activities. We include it as a separate topic, however, simply to underscore an area that is too often ignored by analysts and decision makers when disability policies are being discussed. It is in no way a frivolous aspect of life. There are significant ties to quality of life and mental health status, as well as important symbolic issues relating to equality of access and civil rights of disabled people.

One obvious point is that cities in general have more cultural and perhaps at least as many recreational activities as nonurban areas, and these are generally in buildings or structures of some type, leading to issues of accessible buildings and transportation discussed above. Mobility, however, is only one of several factors that must be considered. Activities must be known. For example, this may require putting information about concerts on computerized bulletin boards or issuing braille announcements.

Even if disabled people are able to get to and enter buildings or areas of recreation and culture, they may not be able adequately to take advantage of them. Theaters and concert halls are usually not equipped with technologies to allow use by disabled people (e.g., for hearing-impaired people rooms can be fitted with induction loop systems, FM broadcast systems, or infrared sound systems), although it is likely that the majority of equipped buildings and rooms are in urban areas. The Kennedy Center for the Performing Arts, in Washington, D.C., is an example.

One key to the availability of accessible cultural and recreational activities appears, again, to be large enough concentrations of disabled people to justify the expense. Thus urban areas are more likely to have disabled sports activities (such as Special Olympics), special days for disabled people at zoos and other sites, and even special parks dedicated to accessibility and use by disabled groups. The interaction of urban functions must be kept in mind: Making recreational opportunities available to disabled people means not only making the buildings, parks, and activities usable but also coordinating transit systems, information dissemination, community groups; it may even involve the subsidizing of some activities or individuals.

Special thought may also need to be given to opportunities for inner-city disabled people: There are fewer parks, transportation is less accessible (particularly in its connections to outlying sites), cost is more of a problem, and communication is harder.

INDEPENDENT LIVING

Most of what has been said above has directly or indirectly been related to independent living by disabled people. Independent living as a concept has at least three critical elements: Individuals make their own decisions to the maximum extent possible; they are integrated into the community as much as possible; and they receive a complex of support services in order to maintain independence.

The number of independent living centers (ILCs) is growing, and, naturally enough (again, considering the issues of concentrations of people), they are probably more common in metropolitan areas (this, however, is a subjective perception[1]). ILCs are usually not residential centers; they are focal points where individuals can receive coordinated support services such as device repairs, employment counseling and referrals, transportation, attendant care, fellowship, and assistance in finding funding sources.

The primary urban versus nonurban consideration related to ILCs and independent living in general is that they seem even more valuable in urban areas because of their ability to provide coordination of services and service referrals. Urban areas, after all, have a sometimes confusing array of funding programs, employment programs and training programs, device repair companies, transportation alternatives, health care and rehabilitation organizations, and social opportunities.

Urban policymakers can often make their jobs easier by becoming acquainted and working with ILCs. ILCs are many times oriented to low-cost, low-technology solutions to problems. They know about gaps in services. More important to a city's fiscal stance, they know about duplications of service and inefficient use of funds. The experience of ILCs in Ann Arbor, Berkeley, and San Francisco are good cases to investigate.

DISCUSSION:
THE CITY AND DISABILITY

In doing the research for this chapter we were surprised by the relative lack of attention given to disability policy (especially technology-related policies) in the urban affairs literature we examined. Other than in the area of transportation, there seems to be little discussion of the problems and opportunities related to the interaction of the urban situation, technology, and disabilities. As an extreme example, one source recommended to us was a two-volume set of proceedings from a Belgian conference on human well-being in cities, which covered a wide range of quality-of-life issues in urban living (Laconte, 1976/ 1977). Not one article dealt with disability and the city. Further, in a thirteen-page, single-spaced index there was no reference to "disability," "handicap," or any related term. Examination of the text itself showed the index to be accurate.

This leads to an extremely important conclusion. If the inattention in the literature reflects actual, inadequate attention by urban governments and analysts, as it seems to do, then a significant gap in policy exists, with strong implications for quality of life as well as public resources.

Two issues that have not yet been discussed should be mentioned: the importance of considering demographic changes in the population of elderly people and the role of analytical methods in urban policy on disability.

INCREASED NUMBERS OF ELDERLY PEOPLE

Issues related to disability will become increasingly important to urban governments as the age distribution of the U.S. population continues to shift, with a growing proportion being elderly. That type of shift implies increased numbers of disabled people, people with limitations in their ability to carry out typical functions. The aging process is associated with reduced ability to see, hear, and maintain mobility in a higher percentage of people than for the general population (U.S. Congress, Office of Technology Assessment, 1982). The incidence and prevalence of chronic diseases, such as cancers, heart and circulatory conditions, and arthritis, increase with the age of a population.

Today, about 12 percent of the population is age 65 and over, and that percentage may rise to 20 percent or more by the year

2030. Technological advances in the next 30 years could significantly reduce mortality (death) and morbidity (disease) rates of elderly people, and the proportion of elderly people could rise more dramatically than anticipated. One potential problem posed by an elderly population is the disparity between the fiscal and social contributions by elderly people that policies require (or encourage or allow) and the fiscal and social needs that increased numbers of elderly people will present.

Urban governments, then, will be faced with a heightening of the current problems associated with mobility, accessibility, health and social services, and so forth. This issue will become even trickier if the trend toward suburban populations with high proportions of elderly people continues (primarily due to the large number of couples that moved to, and have stayed in, the suburbs after World War II), with implications for transportation planning, community services, local area budgets, and most other services associated with special responses to disabled and elderly people.

THE ROLE OF ANALYTICAL TECHNIQUES

One of the most disturbing aspects of the disability policy area (or most areas, for that matter) is the lack of research that has gone into the development and validation of appropriate analytical tools. Traditional private-sector tools, such as return-on-investment analysis, are usually inadequate for public policy purposes. The techniques of cost-benefit and cost-effectiveness analysis (CBA and CEA), developed principally for public decision making, have serious methodological and practical limitations (U.S. Congress, Office of Technology Assessment, 1980). They should not, especially when they force a single, bottom-line quantitative result, be the sole or primary determinant of a decision.

Analyses that calculate the "cost per disabled rider" involved in modifying subways and buses or establishing a dial-a-ride service are *not* adequate cost-benefit or cost-effectiveness analyses. They usually amount to nothing more than simplistic estimates of which system has lower direct costs, ignoring many indirect costs and benefits, uncertainties and nonquantifiable factors, some of which may be among the most important considerations. This is in fact a common shortcoming of cost-related analyses (U.S. Congress, Office of Technology Assessment, 1980). For example, the U.S. Congressional Budget Office (CBO) study (1979) did not take into account indirect benefits such as convenience or benefit from using the same transit services as nondisabled people. In fact, the CBO recognized this point and language in the report of the study indicates that they are using "cost per trip as a guide to the cost-effectiveness of each option." Nevertheless, publicity on the CBO study concentrated on the variables used, the hard numbers. As usually happens, the caveats played little role in considerations relating to the report, until brought up by other parties.

A more comprehensive (investment- and environment-oriented) approach to urban disability policy will require that some thought be given to supplementing traditional forms of CBA and CEA with techniques that force the explicit consideration of interaction effects, indirect consequences, and social and value impacts of policies. For example, traditional "bottom-line" forms of analysis have substantial problems in dealing with social impacts and ethical or value questions. Analysts can enhance their ability to deal with such issues by making a purposeful decision not to seek a single, bottom-line estimation of outcomes. Arraying potential results or outcomes, providing qualitative or at least descriptive information on the not easily quantifiable factors involved, and providing estimates of importance or magnitude for nonquantified variables are examples of techniques that might be considered in order to avoid an easier but perhaps less accurate "answer" for analysts to give to decision makers.

CONCLUSION

We hope that this chapter has succeeded in pointing out the importance of planning for disability-related problems and opportunities when formulating urban policies, especially those directly and indirectly related to technology. Disabled people require a higher-than-average share of public resources, but that share is larger than it needs to be. With informed application of technology, the need for income maintenance and other special services can be greatly reduced over time.

Technology most often has its direct impact on the physical environment, but the consequent social, economic, and psychological effects are usually the more important determinants of successful policies. Thus, a city's ability to make successful, need-reducing investments will depend in part on its ability to view itself and the disabled population as parts of complex set of interacting functional environments.

For example, transportation systems should be shaped not only by the goal of mobility itself but also by the goals of employment, recreation, housing, shopping, health care, and other functions. In turn, it will help to shape those other functions. Rehabilitation centers must be coordinated not only with other medically related facilities but also, for example, with employers, transportation planners, housing authorities, recreation officials, and education administrators.

It sounds complex, and it is. But at the same time it is essential if resources are to be effectively spent and invested, if disabled people are to evade the handicapping effects of the city and enter the mainstream of urban life, and if cities and their governments are to avoid socially disruptive gaps and wasteful duplications in public services.

NOTE

1. Kent Hull of Michigan State University has written that the State of Michigan has supported ILCs in urban, academic-oriented areas, but that recently it has begun providing some degree of independent living services in rural areas.

REFERENCES

American Demographics (1982) "Disability." Vol. 4 (July/August): 24.
American Demographics (1982) "How people get to work." Vol. 4 (July/August): 23.

Bowe, F. (1980a) *Access to Transportation*. Washington, DC; American Coalition of Citizens with Disabilities, Inc.

Bowe, F. (1980b) Rehabilitating America: Toward Independence for Disabled and Elderly People. New York: Harper & Row.

Brown, S. (1982) "A case study of evaluation research in the legislative process: public transportation for the handicapped," in L. Saxe and D. Koretz (eds.) *New Directions for Program Evaluation: Making Evaluation Research Useful to Congress*, No. 14. San Francisco: Jossey-Bass.

Buchanan, J. and M. Chamberlain (1978) "Mobility of the disabled in an urban environment." *International Journal of Rehabilitation Research* 1: 565.

Butler, J. et al. (1981) *Health Care Expenditures for Children with Chronic Disabilities*. San Francisco: University of California, San Francisco, Institute for Health Policy Studies. (unpublished)

Cannon, D. and F. Rainbow (1980) *Full Mobility: Counting the Costs of the Alternatives*. Washington, DC; American Coalition of Citizens with Disabilities, Inc.

Chollet, D. (1979) *A Cost-Benefit Analysis of Accessibility* [prepared for the U.S. Department of Housing and Urban Development]. Washington, DC: U.S. Government Printing Office.

Cohen, H., D. Kligler, and J. Eisler (1981) "Urban-community care for the developmentally disabled." *American Journal of Mental Deficiency* 86: 110-111.

Cornehls, J. and D. Taebel (1976) "The outsiders and urban transportation." *Social Science Journal* 13 (April): 61-73.

Donabedian, A., S. Axelrod, and L. Wyszewianski [eds.] (1980) *Medical Care Chartbook, Seventh Edition*. Ann Arbor: AUPHA Press and Department of Medical Care Organization, School of Public Health, University of Michigan.

Hulek, A. (1969) "Vocational rehabilitation of the disabled resident in rural areas." *Rehabilitation Literature* 30 (September): 258-262.

Koshel, J. and C. Granger (1978) "Rehabilitation terminology: who is severely disabled?" *Rehabilitation Literature* 39 (April): 102-106.

Laconte, P. [ed.] (1976/1977) *The Environment of Human Settlements: Human Well-Being in Cities*, Vols. 1, 2. Oxford: Pergamon.

LaRocca, J. and J. Turem (1980) *The Application of Technological Developments to Physically Disabled People*. Washington, DC: Urban Institute.

Liang, M. et. al. (1981) "Evaluation of a pilot program for rheumatic disability in an urban community." *Arthritis and Rheumatism* 24 (July): 937-943.

MacGuffie, R. et al. (1969) "Rural rehabilitation clients' plans for migration." *Rehabilitation Counseling Bulletin* 13 (September): 44-48.

Montan, K. (1969) "A better urban environment for people with visual impairment." *Rehabilitation Literature* 30 (January): 14-15.

Montgomery, J. D. (1974) *Technology and Civic Life*. Cambridge, MA: MIT Press.

National Arthritis Advisory Board (1979) *Policy and Chronic Disease*. NIH Publication 79-1896. Bethesda, MD: National Institutes of Health.

National Governors' Association, Center for Policy Research (1980) *The Role of the Governor in the Implementation of Programs for the Handicapped*. Washington, DC: National Governors' Association. (transmittal letter dated January)

Poster, T. (1982) "Federal transportation policy for the elderly and handicapped: responsive to real needs?" *Public Administration Review* 42 (January/February): 6-14.

President's Commission for a National Agenda for the Eighties (1980) *Urban America in the Eighties: Report of the Panel on Policies and Prospects for*

Metropolitan and Nonmetropolitan America. Washington, DC: U.S. Government Printing Office.

President's Committee on Employment of the Handicapped (1978) *National Health Care Policies for the Handicapped: A Report to the President by the National Health Care Policies for the Handicapped Working Group.* Washington, DC: U.S. Government Printing Office.

Robey, B. (1982) "Profile: older Americans." *American Demographics* 4 (June): 40-41.

Sigelman, D. et al. (1977) *Human Rehabilitation Techniques: A Technology Assessment,* Vol. 2. Washington, DC: National Science Foundation.

Stedman, D. (1977) "Service delivery systems," in *The White House Conference on Handicapped Individuals, Volume I: Awareness Papers.* Washington, DC: U.S. Government Printing Office.

Stern, V. and M. Redden (1982) *Technology for Independent Living: Proceedings of the 1980 Workshops on Science and Technology for the Handicapped.* Washington, DC: American Association for the Advancement of Science.

Tinker, I. and M. Buvinic [eds.] (1977) *The Many Facets of Human Settlements: Science and Society.* Papers Sponsored by the American Association for the Advancement of Science. New York: Pergamon.

U.S. Congress, Office of Technology Assessment (1980) *The Implications of Cost-Effectiveness Analysis of Medical Technology.* Washington, DC: U.S. Government Printing Office. (Stock No. 052-003-00765-7)

U.S. Congress, Office of Technology Assessment (1982) *Technology and Handicapped People.* Washington, DC: U.S. Government Printing Office. (Stock No. 052-003-00874-2)

U.S. Congress, Office of Technology Assessment (forthcoming) *Medical Technology under Proposals to Increase Competition in Health Care.* Washington, DC: U.S. Government Printing Office.

U.S. Congressional Budget Office (1979) *Urban Transportation for Handicapped Persons: Alternative Federal Approaches.* Washington, DC: U.S. Government Printing Office.

U.S. Department of Health, Education and Welfare, Public Health Service (1976) *Summary and Critique of Available Data on the Prevalence and Economic and Social Costs of Visual Disorders and Disabilities* (prepared by Westat, Inc.). Bethesda, MD: National Institutes of Health.

U.S. Department of Health, Education and Welfare, Office of Human Development Services, Office for Handicapped Individuals (1979a) *Digest of Data on Persons with Disabilities.* Washington, DC; Author.

U.S. Department of Health, Education and Welfare, Public Health Service (1979b) *The Characteristics of Persons Served by the Federally Funded Community Mental Health Centers Program, 1974.* NIMH Series A, No. 20. Rockville, MD: Alcohol, Drug Abuse, and Mental Health Administration.

U.S. Department of Health and Human Services, Social Security Administration, Office of Policy (1980) *Work Disability in the United States: A Chartbook.* SSA Publication No. 13-11978. Washington, DC: U.S. Government Printing Office.

U.S. Department of Health and Human Services, Public Health Service, Office of Health Research, Statistics, and Technology (1981) *Health: United States, 1981.* DHHS Publication (PHS) 82-1232. Hyattsville, MD: Author.

Part II

**The Making and Unmaking
of Health Policy**

Health Care in America:
A Political Perspective

HENRY J. SCHMANDT
GEORGE D. WENDEL

☐ THE HEALTH DELIVERY SYSTEM in the United States is as much a product of politics as of medical technology and science. It abounds with policy issues relating to the quantity and quality of services, the institutional structures through which they are provided, and the regulations that govern their delivery. It involves political choices that determine the share of public resources allocated to health care as against other social needs, the distribution of medical benefits and costs, the supply of health professionals, and the priorities of medical research. Political considerations, moreover, are not limited to national policy issues. They also enter into decisions at the state and local levels in a wide variety of health-related matters, such as facility expansion and licensing, development of ambulatory care centers, Medicaid reimbursement, and licensing of health professionals.

Health care represents a giant industry that accounts for 10 percent of the country's gross national product, entails annual expenditures of nearly $300 billion, and provides employment for more than 3 million people.[1] Until very recently few studies in the health field paid more than passing attention to the politics associated with this huge sector of the service industry.[2] Political scientists evidenced little interest, and researchers in other disciplines, including economics and sociology, concerned themselves mainly with nonpolitical aspects of the system. Today such an observation would be less accurate. During the last decade studies dealing directly with the formulation and implementation of health policies began to appear (e.g., Hyman, 1973; Marmor, 1973; Austin, 1975; Feder, 1977; Brown, 1978; Poen, 1979; Thompson, 1980); and in 1976 a professional periodical emphasizing the politics of medical delivery, *Journal of Health Politics, Policy and Law,* was launched. Despite these promising beginnings, however, health politics remains largely an uncharted field, with communication between political scientists and the realm of medicine still tenuous.[3]

This article takes a broad look at the health care system from a political perspective. It reviews the involvement of government in the delivery of medical services, the issues associated with this involvement, the proposals to change the existing system, and the importance of the nonmedical aspects of the health industry to the nation's cities. It focuses particularly on how health politics functions in a setting in which cost containment is the battle cry, the marketplace the magic word, and governmental decentralization the prevalent orthodoxy. What emerges, not surprisingly, is a picture of policy formulation and execution that is characterized more by political advocacy than by the rational model of decision making.

FROM PRIVATE TO PUBLIC RESPONSIBILITY

When political scientists look at the health delivery sector, they see an enormously complex system (or nonsystem, as some satirically dub it) with a perplexing pattern of relationships, a confusing mixture of public and private suppliers, and a multiplicity of actors with varying degrees of power and influence. They are struck by the highly fragmented character of the system, its almost total lack of coordination, its low level of public accountability, and its financing methods that act as disincentives to efficient and economical operation. These attributes are well recognized by health professionals and academic observers of the field, but they are a source of wonderment to political scientists taking close note of health delivery for the first time.

The emergence of health care as a public or political responsibility is in fact a relatively recent phenomenon. Prior to the Industrial Revolution and its attendant urbanization, the delivery of medical services was almost entirely a matter of private-sector concern. As industrialization proceeded, however, the need for public intervention became increasingly apparent. The realization on the part of the elite that a healthy working class is in the interest of a capitalistic economy was a major factor in stimulating action. The initial involvement of government came in the form of sanitation measures and other public health-type activities (Rosen, 1958). These regulatory functions were followed by health insurance plans designed to assure the provision of medical services to wage earners. Still later, by the middle of the present century, the notion that health care is a basic right of all citizens and an obligation of government became common in most nations of the Western world (Leichter, 1979).

Germany was the first country to adopt a national health insurance program. Its Sickness Insurance Law of 1883 required all workers to participate in a health insurance system financed jointly by employer and employee contributions. England followed in 1911 with a program modeled after the German plan. In the United States compulsory health insurance did not become a major national issue until after World War II. The Great Depression of the 1930s had demonstrated the inability of local and state authorities to cope with the socioeconomic difficulties generated by an industrial society, and the New Deal legislation had broken the tradition of nonintervention by the federal government in social problems. These developments set the change for a greatly enlarged governmental presence in the health care sector.

Franklin D. Roosevelt had considered the inclusion of a health insurance component in the Social Security Act of 1935 but had abandoned the idea for fear that it would jeopardize passage of other measures on his political agenda. His successor, Harry S Truman, sought to add to the social accomplishments of the New Deal by pushing vigorously for

216   CITIES AND SICKNESS

compulsory national health insurance. Under the program advanced by his administration, hospital and medical coverage was to be provided for all segments of the population and financed through a 4 percent increase in the social security tax. When it became apparent that such a broad measure was not politically marketable, the proposal was modified to restrict coverage to the elderly and exclude the provision for physicians' fees. It was this scaled-down version, known as Medicare, that became the focal point of health insurance advocacy in Congress over the next decade.

Medicare finally became a reality following the Democratic landslide of 1964 that kept Lyndon Johnson in the White House and brought a large cadre of liberal Democrats to Congress. In the political bargaining and maneuvering that took place after the election, the Medicare bill was broadened to include a voluntary or elective insurance plan for the elderly covering the fees of physicians.[4] At the same time a companion measure, Medicaid, was enacted authorizing federal matching funds to the states to help pay the medical costs of indigent persons, regardless of their age. (Medicaid represented an expansion of the Kerr-Mills legislation of 1960, which provided for federal medical aid to the elderly poor only.)

The long battle over Medicare reflected the conflict between the public welfare and insurance approaches to the role of government in the provision of health services. Under the welfare concept, public assistance for medical purposes is limited to the indigent ill, a responsibility that in the United States has traditionally rested with private charity and local government. Under the insurance approach, the means test for aid eligibility is eliminated, and all covered participants, irrespective of income, are accorded equal access to health services. Medicare represented a victory — albeit a partial one because of its limited coverage — for the insurance model. By tying hospital care into the social security system and requiring monthly premium payments from those who elected physician coverage, it conveyed a sense of entitlement — the benefits are considered as a matter of "earned right" rather than a dole — and made the program less vulnerable to adverse political action.

The passage of Medicare and Medicaid did not put an end to the health insurance issue. Although the provisions for the elderly and indigent had lessened the urgency for a program of universal coverage, the escalating costs of hospitalization and medical treatment served to sustain strong public interest in such a measure. By the early 1970s most observers were predicting that some form of national health insurance would be enacted before the end of the decade. Great Britain had already taken this step in 1946 when it passed the National Health Service Act. The program made comprehensive medical care available to every British resident and shifted financing of the costs from the contributory insurance scheme of the earlier health measure to general taxation.

Whatever the ideological predispositions of the American people toward national health insurance, the movement to extend coverage to all citizens has fallen casualty to soaring medical costs, economic recession, the tax revolt, and the "curb government" syndrome. The actions of President Carter reflected these developments. During the 1976 campaign he repeatedly promised his support for a national health insurance plan. After taking office he qualified this pledge, stating that the control of health costs was a necessary first step before any such program could be put into operation. By 1979 he was saying that the nation could not afford more than gradually phased-in changes in the existing system. Since the advent of the Reagan administration, the expansion of health care coverage through governmental action has become a political nonissue, with debate now centering on cost containment and reduction of benefits under the Medicare and Medicaid programs.[5]

A PRIVATE OR COLLECTIVE GOOD?

An overview of health politics can usefully begin by placing medical care within the classification framework that is commonly employed in the continuing debate over government's role as a service provider. According to this schema, goods and services have certain intrinsic characteristics that enable them to be categorized as either private or collective (public). The first are consumed by individuals in varying amounts, such as water and mass transit, and they can be withheld from persons who do not pay. The second are jointly consumed, such as national defense and police protection, and it is not possible to exclude their benefits from those who do not contribute to their financing. They are thus subject to "freeloading" by individual consumers. Many analysts, particularly economists, employ this distinction as a theoretical basis for determining the "most efficient" division of service responsibilities between the private and public sectors. The classification is also invoked for ideological purposes by political conservatives who argue that one of the fundamental reasons governments exist is to provide services that are intrinsically collective, as distinguished from private.

There is general recognition that some services defined as private have large positive externalities or spillover benefits (or costs). Education, for example, is available in the private sector, and it can readily be denied to those who do not pay the required tuition. However, not only the individual consumer benefits from the service but the entire society also gains, because of an educated citizenry. The American political system treats a number of services of this kind as collective goods (some social scientists refer to them as "merit goods") and makes them available to all people through the public budget. Conservatives are wont to complain that the burgeoning growth of government in recent decades has resulted from redefining private goods as collective or merit goods.

Health care, to a greater extent than most other service areas, has both private and collective features. Some public health activities, such as air pollution control and food and drug regulation, are pure collective goods (no one can be excluded from their benefits). Others are of mixed character. Immunization, to cite one example, is individually consumed and subject to exclusion, but it benefits the community as a whole by reducing the risk of contagion. The most disagreement occurs over the classification of medical treatment and hospitalization, the two largest components of the health care system. Theoretically they are private goods: They are individually consumed, supplied predominantly by the private sector, and of such nature that they can be denied to nonpayers. Observers who view them in this light maintain that their quantity and quality are matters for determination by the marketplace, with government's involvement limited largely to assisting people who lack the means to pay. Health advocates, on the other hand, argue that these services are merit goods in which price and market competition have little place. In their view, medical care is of such value to society (because of the major importance of individual health to the social and economic well-being of the community) that government should intervene to guarantee all persons, regardless of income or age, equitable access to the delivery system.

Health care in the United States is currently treated as a collective or merit good in the case of the elderly (through Medicare) but as a private good with respect to the majority of the population. This division on the basis of age obviously has nothing to do with the nature of the service but is a product of politics (the elderly have wide popular acceptance as a "worthy" or "deserving" group). Clearly, the determination of what goods warrant public production or subsidy rests more on popular preferences and political realities than on any set of analytically derived categories. Nevertheless, the continued appearance of the private-collective classification in the debate over government's role in the delivery of services cannot be disregarded. The conceptual framework it offers is frequently used to defend, or lend "intellectual credibility" to, positions that are grounded essentially in politics or ideology. An example is the high praise given by Reagan administration officials to a recently published book entitled *Privatizing the Public Sector: How to Shrink Government*. Its author, E. S. Savas (1982), employs the "nature of goods" framework to support his argument for reducing the scope of government, including its involvement in health care delivery.

While many advocates of "load shedding," as Savas refers to the curtailment of public-sector activity, continue to draw heavily on the conventional private-collective distinction for theoretical legitimation, proponents of increased government intervention in the medical field have enlarged their claim of health care as a collective or merit good to include the notion of "right." Senator Edward Kennedy, long a leading supporter of universal health coverage, gave expression to this claim in declaring that "all Americans have come to expect health care as one of the basic rights of citizenship." This designation of right, discounting the rhetoric associated with it, is subject to several interpretations. Politically, it may be construed as reflecting a broad social consensus that all citizens are entitled as a matter of justice and equity to approximately the same level of medical care, and it is therefore government's responsibility to see that they receive it. Philosophically, it may be regarded as an ethical imperative that society is morally obligated to implement whether wide political agreement exists or not. Legally, it may be taken to mean that health care is a judicially enforceable prerogative of citizenship, a contention advanced by some of the more ardent supporters of national health insurance.

For the last interpretation to have validity, the right to medical services must be shown to have a constitutional basis. Although the issue has not been adjudicated as to health care, a similar claim for education as a fundamental right has been denied by the Supreme Court. In a 1973 decision involving inequities in the financing of public schools, the court ruled that "while education is one of the most important services performed by the state, it is not among the rights afforded explicit or implicit protection under the federal constitution."[6] Plaintiffs had argued that even though education is not mentioned in the nation's basic law, it is nonetheless a fundamental personal right because it is essential to the effective exercise of First Amendment freedoms. A similar argument made in the case of health care would undoubtedly meet the same fate as the education claim (Carey, 1974).

THE ROLE OF GOVERNMENT

Aside from the question of classification, the nature and extent of public-sector involvement in the health field is difficult to describe, much less understand. In a delivery system in which the regulated may also be the regulators and in which third parties rather than consumers pay most of the bill, identifying and sorting out the resulting relationships is a complex task.[7] Basically, government plays three major roles in health care: third-party payer, regulator, and provider. All levels of the political structure serve in these capacities, although the extent to which they do so varies. The federal government is the principal public actor, supplying almost 30 percent of the total amount spent annually on medical care in the United States. It is most heavily involved as a third-party payer, a role that accounts for much of the dramatic rise in public-sector activity in the health area over the past two decades. In addition, it performs a broad range of regulatory functions (setting standards for water and air purity, monitoring product safety, and policing food and drug products are examples). To a lesser degree it is also a service provider (operating veterans hospitals, funding medical education, and sponsoring medical research).

The states play similar roles in the health field. They are involved as third-party payers, a responsibility they share with the federal government in the Medicaid program. They exercise substantial regulatory powers (including supervision of nursing homes, certification of need for health facility expansion, and licensing of medical practitioners), and they are direct suppliers of health services (operation of mental institutions and acute care hospitals and training of medical professionals in state universities).

Unlike the other two levels where the role of direct provider is subordinate in scale to that of payer and regulator, local government's chief task in health delivery is the provision of services (operating public hospitals and neighborhood health centers and performing a wide spectrum of public health functions, from screening schoolchildren for dental needs to protecting the community against communicable diseases). Its regulatory responsibilities are limited primarily to sanitation-type activities, such as licensing restaurants and enforcing minimum housing standards. The role of third-party payer is applicable at this governmental level only in those cities and

222 CITIES AND SICKNESS

counties that purchase health care for the medically indigent from private hospitals or that are required by state law to contribute to Medicaid payments.

The diffuse nature of public-sector involvement in health delivery is reflected in the intricate network of relationships that exists among governments and between government and the medical industry. These interrelations are poorly defined and a frequent source of confusion. The federal government occupies a dominant position in the mosaic because of its role as major financier, but it must rely on the states and localities as well as the private sector to carry out its programs. The formal pattern of organizational interaction between political levels, moreover, is not consistent. In some programs the federal government channels funds directly to city and county authorities; in others it subsidizes locally administered projects through the states; and in still others it bypasses both state and local units to deal directly with private providers (as in the case of financial assistance to medical schools and health maintenance organizations).

State-local relations in health matters are also numerous and complicated. The states serve not only as conduits for federal aid to their political subdivisions but also as suppliers of limited financial assistance drawn from their own revenue sources. The degree of interaction between the state and local levels is increasing as the federal government moves from categorical aids to the use of block grants. As this shift continues, it will intensify relations between state authorities and local officials and compel the latter — many of whom are long accustomed to dealing directly with federal agencies — to cultivate closer ties with state grant administrators and legislators.

The health care budgets of the large central cities illustrate the complex pattern of governmental involvement as well as the fiscal dependence of the localities on higher levels of the political system. As they pointedly show, communities must rely on a variety of governmental sources other than local revenue to finance their health care services. In St. Louis, as an example, the city spent a total of $65.8 million for health-related purposes in fiscal 1980. Less than one-third of the money came

TOTAL REVENUE - $65.8 MILLION

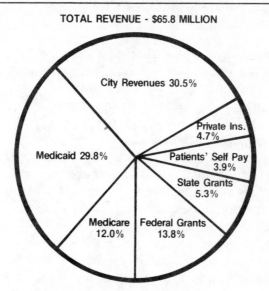

SOURCE: Based on Comptroller's Report, city of St. Louis, 1980.

Figure 7.1   Revenue Sources, Department of Health and Hospitals, City of St. Louis,
Fiscal Year 1980

from the municipality's general revenue fund, the remainder from outside sources. Medicaid payments from the state provided 30 percent of the total, Medicare 12 percent, and federal and state grants almost one-fifth. The combined receipts from the out-of-pocket payments by patients and from their private health insurance, such as Blue Cross, constituted less than 9 percent of the revenue package (Figure 7.1).

Approximately 80 percent of the health budget of the city of St. Louis goes to its municipally owned hospitals. Medicaid payments alone account for almost 40 percent of their revenue. This heavy reliance on Medicaid patients to sustain their operation is an indication of the precarious condition in which public hospitals in St. Louis and other communities now find themselves. Cutbacks in Medicaid reimbursement, along with more stringent requirements for determining eligibility, have contributed to the plight of such hospitals by narrowing their subsidies and making them the providers of last resort for an increasing number of nonpaying, uninsured patients.

THE PATTERN OF POWER

The power configuration in the health delivery system is as diffuse as the organizational structure that encompasses it. Most observers who give attention to health politics draw at least nominally on one or another version of interest-group theory to embellish their analyses of the policymaking process (Greenstone, 1975; Marmor and Marone, 1980). Some follow the pluralist model, which views policy choices as the product of mutual accommodations among concerned and competing interests. This model, in its more benign form, assumes that all legitimate groups can organize and make their voice heard at some point in the decisional process (Dahl, 1964). Others see the health care system and its pattern of power in structural or elitist terms (Alford, 1975). They reject the pluralist notion of government policy as an amalgam of the desires of competing groups, contending that such a view overstates the facility with which interests can organize and understates the disproportionate degree of resources among contesting parties. According to their version, health policy is dominated by major interests that are well served by the existing class structure of society, with their power and resources safely embedded in law, custom, and professional legitimacy.

Those who espouse the pluralist model properly conceive the health care system as encompassing a multiplicity of interacting groups and institutions, each with its own priorities and objectives. They fail in general, however, to distinguish clearly between group participation and group influence, and their treatment of interest representation is overly sanguine. The structuralists, on the other hand, correctly emphasize the wide disparity in influence resources among the various concerned interests and the power advantages that some groups — the medical profession in particular — enjoy by virtue of their high status in society. However, they tend to overstate the subservience of the political process to provider influence (witness the adoption of Medicare and Medicaid over the opposition of the American Medical Association) and to ignore the major role that government now plays in the financing and delivery of health care.[8] They also fail to give sufficient recognition to cultural and attitudinal changes that alter the power potential of various interest groups (higher public expectations of medical services in the post-World War II period significantly strengthened the influence of health care advocates). The medical profession may remain the dominant actor in health policymaking, as the structuralists contend, but its power is being increasingly challenged by other groups within and outside the health field and by politicians and government bureaucrats (Weller, 1977).

Neither the pluralist nor the structuralist model offers more than nominal assistance to an assessment of power in service delivery systems. Both are deficient in that they do not specify the circumstances under which, and the kinds of issues in which, the various special interests will or will not be influential (Hayes, 1978). They fail, moreover, to provide an analytical framework that takes account of the fact that policymaking authority over certain issues may be divided among more than one level of the political system. The persistent controversy over abortion and the joint federal-state arrangements in the Medicaid program are illustrations of the complex, multitier division of power and decision making in health-related matters.

Each governmental level in the American polity deals with health issues, sometimes singly and at times in conjunction with its counterparts in other tiers. Prior to the late 1970s the federal government was concerned primarily with redistributive policies and programs, such as Medicare and Medicaid, that were designed to bring about a wider and more equitable apportionment of medical benefits. Since that time the emphasis has shifted to policies aimed at halting the rapid rise in health costs. When redistributive issues were paramount, the various segments of the health industry were joined in their opposition to such measures by conservative organizations outside the medical sector, including the National Association of Manufacturers and the U.S. Chamber of Commerce. When the focus turned to cost containment, however, ideological ties tended to give way to self-interest, with the outside groups becoming less supportive of the health establishment's opposition to governmental action against medical inflation. Many large corporations, for example, are in favor of planning controls over hospital expansion, and many have also indicated that they no longer regard the fee-for-service practice as the most efficient way of delivering medical care.[9]

At the local tier the emphasis is on issues associated with the direct provision of health services, such as the operation of ambulatory care facilities and the maintenance of public hospitals. These matters tend to activate a constellation of interests different from those at the national level. Health providers remain the most prominent figures in the power pattern, but they are not always in agreement, as in disputes over hospital expansion, and they are more open to challenge from community-based organizations. Until recent years, business and civic leaders also played a greater role in the decisional process at the municipal and county levels because hospitals had to rely on local philanthropy to finance a portion of their capital needs. (Most hospitals now finance new construction and the acquisition of major equipment through the issuance of bonds.[10]) And because most health issues that fall within the purview of local government have little ideological content (abortion is a major exception), the medical establishment has fewer opportunities to mobilize public support on its behalf by invoking the traditional value symbols, such as "keep politics out of medicine."

The most important changes in decision-making authority have been occurring at the state level. The adoption of Medicaid in 1965 substantially enlarged the involvement of state governments in the health field by placing responsibility for administering (and partially financing) the program in their hands. More recently — since the advent of the Reagan presidency — their powers have been further expanded by the shift from categorical aids to block grants. Under the budget reconciliation bill of 1981, twenty-one programs of health assistance to local governments were consolidated into four block grants to the states. The transfer lowered the allocation decision process — both as to which individual programs will be funded and which localities will receive the money — from the federal government to the states. (The block grant concept requires the states to use the funds for the general purposes stated by Congress but normally does not require them to continue any specific program in the group.)

Moving allocational authority from the national bureaucracy to the states lodges it in a new political setting. Because little has been done to examine health care policymaking in state government, we can only speculate as to whether the various interests, such as the hospitals and consumer groups, exercise the same degree of influence at this level as they do at the national. For the same reason we can only conjecture whether the relative power of the state health bureaucracies is commensurate with that of their federal counterparts.

The medical profession has obviously fared well at the hands of the states, as evidenced by its domination of the regulatory boards that license physicians and the more than necessary restrictions commonly imposed on alternative providers, such as nurse practitioners, midwives, and paramedics. In relative terms, its influence over the regulatory process at the national level has not been as great. Primarily for this reason, the AMA has been waging an all-out campaign to secure the exemption of licensed professionals from the jurisdiction of the Federal Trade Commission, arguing that the states should be permitted to regulate the professions without federal interference.[11]

An important question relative to the power constellation is whether the shift from the federally administered categoricals to the state-administered block grants will affect the continuance of individual programs and the distribution of the funds among local units. At this early stage it appears that most programs will survive, but at a lower funding level and with some modification in their proportionate share of the available aid. One reason for not anticipating radical change is that Congress (as noted in the next section) has been reluctant to give the states *carte blanche* authority over the grants. Another is that vested interests have developed around the categoricals that are politically difficult for state policymakers to ignore. The programs most vulnerable to cutbacks or elimination are those that redistribute tax revenues to benefit low-income citizens and those with narrow constituencies or limited applicability, such as the prevention of poisoning from lead-based paint (Nathan et at., 1983).

As to the question of who gets what among the local units, the initial experience in Missouri and several other states indicates that rural areas and small towns, many of which had not shared in the categoricals, will fare better under the block grants. How widely the aid is dispersed, however, will depend in part on the type of distribution process that is employed. States allot programs funds in essentially two ways: according to formulae adopted by legislative or administrative action and on a discretionary or competitive basis. Under the first method the money is spread throughout the state, generally to the disadvantage of the cities and urban counties. Under the second, competition among applicants encourages better-planned projects and affords grant administrators opportunity to concentrate funds in areas where their potential impact is greatest. This method usually favors the larger cities and counties because of their technical expertise and grantsmanship skills. Both procedures are in common use, with political factors rather than logic determining the choice in the case of each individual program. The result is a mixed distributional pattern that differs from state to state, with the formula basis tending to be more prevalent.

REFORM AND POLITICAL REALITY

The two central issues in health care politics today are decentralization of federal control over the grant programs and cost containment. Both relate to the same objective: the achievement of greater efficiency in the delivery of health services. The first involves the domestic battle plan of the Reagan administration to shift responsibility for social programs to the states and localities. The second, by far the more critical issue because of escalating health costs, revolves around two sets of proposals, one based on marketplace principles, the other on governmental regulation. The policy inconsistencies that have resulted from efforts to resolve these issues are simply a reflection of political realities.

The current version of new federalism calls for block grants to the states with no strings attached. Congress has accepted the concept in principle but has been reluctant to relinquish full control over the way in which the funds are spent. Members are particularly concerned with protecting their favorite programs from termination or severe curtailment by state authorities. This reluctance finds expression in Congress's retention of certain programs as categorical aids and its attachment of various conditions to the block grants.

The controversial family planning program illustrates the political interplay over grant administration. The 1981 budget reconciliation bill continued it as a categorical aid despite Reagan administration efforts to have it consolidated into a block grant. Supporters of the program feared that such a transfer would be tantamount to killing it in some states. To win backing for its retention as a categorical, they accepted a proposal from conservative Republicans to include funds "for discouraging sexual activity among teenagers." The same sort of political bargaining took place in connection with most of the other health programs slated by the administration for conversion to block grants. As a result, each of the four block grants adopted in 1981 contains conditions on use of the funds. One, in fact (Primary Health Care), consists of only a single program, a congressional ploy intended to ensure its continuation by the states.

The cost-containment picture is even more complex, with deep substantive and philosophical differences separating the market enthusiasts from advocates of governmental intervention. Those who regard the marketplace as the most effective regulator of costs seek to place health care delivery on a competitive basis and eliminate existing disincentives to efficient operation. The proposals they offer fall into three broad categories: deregulation, promotion of competition, and consumer cost-awareness.

Proposals in the first group are designed to remove what their sponsors see as obstacles to the free play of the market. Among other changes, they call for the termination of the health systems agencies (HSAs) that were created to perform health care planning on a metropolitan or regional basis, and the professional standards review organizations (PSROs) that provide peer review of hospital admissions.[12] The medical establishment has welcomed efforts to remove governmental restrictions on its operations, but it ignores the fact that its own monopolistic practices have served as effective constraints on marketplace competition. It was not until the late 1970s, for example, that the American Medical Association's ban on physician advertising was successfully challenged by the Federal Trade Commission; and only recently did the Supreme Court rule that a local medical society had violated antitrust laws by engaging in price fixing.[13]

Efforts to promote competition in the health field were initiated prior to the Reagan administration. Most proposals directed at this objective look to some form of prepaid group practice as an alternative to the provision of care by solo practitioners on a fee-for-service basis. The favored vehicle is the health maintenance organization (HMO), which offers comprehensive medical services to its members for a fixed annual fee.[14] Because of its financing features as well as the increased efficiency made possible by group practice, this mode of delivery is presumed to be more cost-effective than the traditional method of supplying medical services. As its supporters point out, it gives providers an economic incentive to utilize preventive and outpatient treatment in order to keep their costs down.

HMO prototypes have existed since the early decades of the present century, but it was not until after 1970 that this form of organizing medical services began to attract the attention of governmental policymakers as a possible means of controlling costs. To encourage the establishment of such groups, Congress in 1973 authorized grants to help finance their capital needs and initial operations. The legislation also requires employers who pay their workers' health insurance premiums to offer an HMO option if such a federally certified facility is available in the area. The law was passed over the opposition of the AMA, which argued that such aid subsidizes one form of medical practice over another.

The HMOs have in the main functioned well (Harrison and Kimberly, 1982), but they have had relatively little impact in promoting competition. This failure is partially attributable to provisions of the law itself. In its final form the HMO bill represented an amalgamation of two diverse philosophical positions: that of the marketplace advocates, who looked on it in terms of efficiency, and that of the social reformers, who viewed it as a way of giving low-income households better access to the health care system.[15] To qualify for federal aid, HMOs had to meet certain socially related requirements and standards that virtually priced their benefit package out of the market. In any case, the federal HMO subsidy program is now being phased out, partly due to its own timetable as a temporary support for new HMOs and partly due to the Reagan administration's contention that HMOs should be able to stand on their own feet. Even without federal aid for the future, the 269 HMOs with 11.6 million enrollees at the end of 1982 seem to constitute a significant alternative form of medical care, though not the fundamental transformation their most optimistic advocates predicted (Demkovich, 1983a).

Proposals in the third category — to make health consumers more cost conscious — are based on the rationale that people use more medical care than they need because they are shielded from the full costs by private or public insurance (Kallstrom, 1978). Increasing their awareness of these outlays, so the reasoning goes, will lead to competition among health care providers and compel physicians to form group practice organizations in order to operate more efficiently. Proposals designed to accomplish this objective range from requiring greater cost sharing by patients to using tax incentives to encourage less comprehensive forms of health insurance coverage for workers.

The most effective way of raising the level of cost consciousness is to increase substantially the proportion of the medical bill paid by the consumer. (Currently, less than 30 percent of all expenditures for personal health services represent out-of-pocket payments by patients.) Aware that such a frontal assault on the existing payment structure would have little political appeal, Congress nevertheless, in 1981 and 1983, increased costs to Medicare beneficiaries in order to reduce federal expenditures.[16] Most proposals to shift costs to the

consumer call for the removal of at least part of the favorable
tax treatment now given to group health insurance premiums
that companies pay for their workers as a fringe benefit. In
essence they would limit both the size of the deduction that
firms may claim as a business expense for such payments and
the amount of the premium that is nontaxable as income to the
employees. These limitations are intended to encourage firms
to offer and workers to choose cheaper plans that are less
extensive in their coverage. The logic may be sound, but it is
extremely doubtful that proposals of this nature can attract a
viable constituency for their implementation. Labor unions are
strongly opposed to such action as an intrusion into the traditional collective bargaining process; business is wary of the
administrative costs associated with this approach; and many
conservatives look askance at using the federal tax code as an
instrument for changing social policy.

From a political perspective, the market-directed proposals
may be faulted for failing to take into account cultural factors
and popular attitudes. As a society, Americans place a high
premium on health (in a recent poll, over 80 percent of the
respondents ranked "having good physical health" first when
asked what they valued most). They also have a strong personal preference for the traditional form of medical care and are
averse to bureaucratic medicine with physicians working in
organizational settings (a 1980 Harris poll showed that approximately three of every five eligible nonmembers were not interested in joining an HMO). Market solutions do not address
the question of these value preferences. Nor do they give
sufficient recognition to the production side of the coin. Health
providers have little liking for the competitive approach,
much preferring the relative security of the regulatory, process — which they have been fairly successful in manipulating
to their advantage — to the uncertainty of the marketplace
(Feldstein, 1979).

The proposals to contain costs through governmental intervention as distinguished from market controls run the gamut
from utility-type regulations to strengthened procedures for
monitoring hospital and surgical utilization. Measures that seek
to curtail costs directly by fixing the price of health care ser-

vices or placing a cap on annual increases in hospital rates, as sought by the Carter administration, had little political acceptability. The approach that received the most support from governmental policymakers in the pre-Reagan period involves the use of planning as a means of containing health expenditures. It is based on two premises: (1) An excess supply of hospital beds and high-technology equipment stimulates their utilization beyond what is medically unnecessary. (2) Planning can "scientifically" determine the need for hospital expansion and health services and provide guidance for the optimal employment of existing facilities.

Beginning with New York in 1964, a series of states enacted certificate of need (CON) laws requiring a determination of community need by a state review board as a prerequisite for the licensing of new health facilities. Congress complemented this approach in 1972 by establishing a capital expenditure review process for hospital expansion known as Section 1122 (an amendment to the Social Security Act). The program was linked to reimbursement provisions rather than licensure. Health institutions that engaged in facility construction without state approval were to be penalized by a partial loss of reimbursement for Medicare and Medicaid patients (Brasfield, 1982).

Implicit in the certificate of need concept is the use of health planning to provide a knowledge base for the review of proposed capital projects. Concerned about the lack of such planning in most states, Congress in 1974 moved to create a two-layer system of health planning agencies. The law required all states to set up a planning body at the state level and health systems agencies (made up of locally selected providers and consumers) at the metropolitan or regional level. It also mandated the states to adopt a CON program congruent with federal guidelines or face the loss of all health grants. HSAs were to perform the local planning and conduct the initial review of proposed new health facilities and major equipment purchases, while the state agencies were to make the final determination and issue or deny the certificates of need.

Implementation of the 1974 act has proceeded slowly. Although more than 200 HSAs have been established, many states have failed to set up the prescribed CON program (Congress in 1979 and again in 1981 postponed pending cutoff of funds to those not in compliance).[17] The meeting of such a requirement may soon be a moot question, because the planning network contemplated by the law now faces almost certain extinction. Consistent with its antiplanning philosophy and its efforts to reduce the regulatory role of the federal government, the Reagan Administration has targeted the program for early termination.[18] The 1981 budget reconciliation bill temporarily prolonged the life of the HSAs by continuing their funding for fiscal 1982, but at a sharply reduced level. Unexpected support for the reprieve came from a handful of Republican conservatives who were hesitant to abandon planning controls before the adoption of pro-competition measures to contain costs. Although the action by Congress saved the program from immediate termination, the severe funding cutback (coupled with a provision in the budget bill that permits governors to abolish HSAs if the state promises to assume their functions) has dealt a serious blow to federal coordination of health planning efforts.[19]

As these various proposals illustrate, policies to curtail costs may be either supply-side or demand-side oriented, the first directed at providers, the second at consumers. Planning controls and the other forms of government regulation fall into the supply-side category. Their purpose is to impose restraints on health service providers in the interest of consumers. Conversely, most market-oriented policies, such as deregulation, tax incentives, and the use of vouchers,[20] reflect demand-side strategies. Their objective is to influence consumer behavior, or more specifically to reduce demand by offering incentives for users to choose cheaper and less inclusive health care insurance. Both demand- and supply-side measures are currently before Congress.

Health analysts are not in agreement on which side of the equation should be pursued. Many economists maintain that the demand side must be the focus of reform efforts, citing it as the fundamental source of health care problems (Joskow, 1981; Steinwald and Sloan, 1981). Others argue that the supply side deserves at least as much attention as its counterpart.

Economist Alain Enthoven (1980), for example, contends that efforts should not be limited to influencing consumers to demand less health care but should also be directed at motivating providers to be more efficient in the use of medical resources. These economic perspectives dominate the national debate over the curbing of health costs, but they must be measured against political realities. Whether policies on one or the other side are pursued will be determined less by economic theory than by the interplay of the political process.

Pro-competition proposals aimed at the demand side are appealing to many policymakers because it is easier to talk about expanding consumer choice than about imposing controls on the influential and well-organized providers. Yet the attainment of a more competitive market cannot, as Enthoven points out, be achieved by demand-side measures alone. Unwarranted supply-side restraints must also be removed, such as the anticompetitive rules promulgated by professional licensing boards and the restrictions on physician participation in corporate practice. It is here that the administration's new federalism thrust of transferring more decision-making power to the states faces a dilemma. Implementing the proposals for an open market in medical service delivery would require greater, not less, federal intervention, because many of the governmental restraints on health care competition are state-imposed (Lippincott and Begun, 1982).

Faced with deficit estimates exceeding $200 billion for the Medicare hospital trust fund by 1995, Congress in March 1983 enacted a supply-side intervention bill of great potential for future medical cost control. Under the measure, hospitals are to be reimbursed for Medicare services to the elderly according to a fixed rate set in advance for each kind of treatment. This new method of payment contrasts with the past system, in which hospitals received Medicare reimbursement "retroactively" on a cost-plus basis. As analysts have long pointed out, the latter method provides no incentive for holding costs down. Most proponents of "prospective" payment go beyond its application to Medicare, urging that the system be imposed across the board to include hospital and physician costs, which are reimbursed by government or private insurers. Whether such expansion takes place will depend in large part on the success of the prospective payment method in slowing the escalation of hospital costs for Medicare's elderly beneficiaries. It will also depend on the reaction of physicians to the new restraints it imposes on hospital costs and medical procedures (Demkovich, 1983b).

If cost-containment efforts, whether of a demand- or a supply-side character, are to have significant impact short of reducing the level or quality of medical care, they must ultimately change fundamental aspects of the health system. These include caring for patients in less expensive facilities than contemporary hospitals and providing treatment through more efficient forms of medical delivery than the individual practitioner. Interventions of this nature run counter to the deeply entrenched public preferences for such traditional features of the health system as hospital-based care and the personal physician-patient relationship. Whether the political process can bring about modifications of these attitudes sufficient to permit basic reforms in the system is highly problematic.

THE HEALTH CARE INDUSTRY
AND THE CITY

The health care industry is important to local communities for reasons that go beyond the medical services it provides. It is a significant generator of employment (badly needed by many central cities and urbanized counties), and its facilities are often critical to urban redevelopment and neighborhood stabilization. In the St. Louis SMSA, to cite a typical example, health services in 1978 involved 66,000 jobs, or about one-third of all nongovernmental service positions in the area, with an annual payroll of $685 million. The wide variation in size of the more than 3,000 establishments in the area employing health service personnel is a further reflection of the system's complexity. Of this number, over two-thirds (largely the offices of private medical practitioners) had less than five employees. At the other end of the continuum were eighty health care establishments with more than 100 workers, the sixteen largest (all hospitals) employing 1,000 or more.

Local government jurisdictions in urban areas commonly compete to attract health care establishments both for the economic benefits associated with them and the medical services they provide. Offices of physicians and dentists contribute to neighborhood and commercial area vitality, and a medical office building is often an important component of a shopping center or planned unit development. Alan Altshuler (1965), in his perceptive book on planning, describes how a proposed new public acute care hospital in St. Paul, Minnesota, in the late 1950s became a vigorously sought-for development prize, with an extensive array of economic interests, political actors, and professional groups competing to influence the locational decision.

Today the question rarely concerns the construction of a new public hospital but rather the closing of an existing one because of rising costs, declining revenues, and reduced utilization (attributable mostly to population losses in the cities and the private hospital alternative made available to the poor by Medicaid coverage). In most instances where public hospital closings have taken place — as in St. Louis, New York, and Philadelphia — the facilities were located in minority-populated neighborhoods and served mainly the residents of those areas. The struggle on the part of minority interests to retain the hospitals has invariably been bitter and complex. In the case of St. Louis, the reopening of such a facility has now become a political issue on which mayoral elections are won or lost.

The importance of the health industry to urban areas, particularly to the older central cities, cannot be emphasized too strongly. Many private hospitals and physicians have followed the population exit to suburbia, leaving the poor and elderly residents of the inner city increasingly dependent on public health facilities. The large university-associated medical complexes, on the other hand, have generally remained in the central municipality, spatially anchored by their huge investment in buildings and facilities. Not only have they stayed, but many of them have also assumed active and leading roles in revitalizing the area of their location. In St. Louis, the medical schools (and their complementary hospitals and institutions) at both St. Louis and Washington universities have established redevelopment corporations to "salvage" and upgrade their surrounding neighborhoods — a form of private urban renewal emphasizing rehabilitation and conservation. As redevelopment entities, they enjoy certain governmental powers delegated to them by law, such as the right of eminent domain. In this role also, they are competitors for community development block grant (CDBG) funds and other government aid for public improvements in their neighborhoods.

Health care facilities of lesser size and scope than general hospitals and medical institutions are also important elements affecting the fortunes of urban neighborhoods and the political processes of cities. The evolution of the maternal and infant care clinics, which began in the 1930s, into the neighborhood health centers of the Great Society in the 1960s brought with it an expansion in the purposes of locality-based facilities. As noted by various analysts (Davis and Schoen, 1978; Davis et al., 1983), these centers came to be viewed as ideal instrumentalities for integrating many of the goals of the War on Poverty and Model Cities, including the devolution of policymaking and citizen involvement. Located as they were in poverty areas, they achieved a measure of decentralization of primary medical services from the ambulatory care facilities of the large public hospital to the neighborhood level. They also provided employment opportunities for locality residents and gave them some degree of policymaking power through membership on the health center boards.

Some of the neighborhood centers were absorbed by the municipal health systems as the New Federalism of the Nixon administration succeeded the Great Society of the Johnson era. Others survived by becoming "freestanding" centers (not part of a general hospital), supported by federal funds and largely uncoordinated with the health systems of the local governments. The freestanding centers, however, face a questionable future. Although they may be a cost-saving alternative to serving the poor in the outpatient clinics and emergency rooms of acute care hospitals, they must now rely for their funding on the primary care block grant to the states. This shift injects an element of uncertainly into the allocational process, in contrast to the more secure funding pattern that existed under the former categorical program administered by the Department of Health and Human Services.

CONCLUSION

The cost crisis has transformed the political landscape in the health care field (Crawford, 1979). The politics of expansion that has characterized the health delivery system during recent years has given way to the politics of cost containment. No longer is the overriding issue one of extending compulsory health insurance coverage to the entire population; it is now one of reducing costs through government regulation or market competition. The Reagan administration and most conservatives opt for the competitive solution. But as economist Lester Thurow (1980) has noted, no group — despite rhetoric to the contrary — is really willing to accept the distributive results of the marketplace. Neither the medical profession nor health consumer advocates favor a competitive delivery system: the former, because it perceives such an approach as threatening the monopolistic practices it has long employed; the latter, because they fear that the health needs of the poor will suffer if left to the private marketplace.

The preoccupation with cost containment has caused the nation to lose sight of the significant improvement that has taken place in the quality and availability of health services over the last two decades. Medical care is not only keeping Americans healthier and alive longer; it is also — thanks to the political process, not the marketplace — reaching a higher proportion of the citizenry than ever before. Although serious inequities continue to exist, virtually all subpopulation groups in the United States, including the poor and racial minorities, have greater access to the health delivery system today than they did just fifteen years ago (Aday et al., 1980). Most observers acknowledge these gains but maintain that the country is paying too high a price for the care it is receiving because of inefficiencies in the delivery system. How these defects can be corrected elicits no agreement among them.

Some critics go beyond the inefficiency problem, contending that the nation is devoting excessive resources to medical services. But given the priority that individuals place on health, it is doubtful that many would be willing to trade lower costs for a lower level of care or make medical choices on the basis of costs. Ironically, people rely on medical treatment to sustain their health, but many tend to behave in ways that contribute to poor health, as evidenced by their refusal to stop smoking, to wear seat belts, to consume less liquor, or to change over-stressful lifestyles. Despite mounting proof of the effectiveness of preventive measures, such as physical fitness activities and the reduction of environmental hazards, the curative philosophy continues to dominate the health field.[21]

Aaron Wildavsky (1979), in his book on policy analysis, lists three generic types of health care systems: public, private, and mixed. The first involves total government coverage ("socialized medicine"); the second represents the market-oriented model (with protection against catastrophic medical costs); and the third is a combined public-private approach (such as now exists in the United States). The mixed model, according to Wildavsky, is the least desirable of the three — it is "bad in all respects" — and is doomed to inevitable failure because it does not impose strict enough discipline on medical costs at either the individual or the collective level. We may well argue, however, that a system that has produced substantial gains in medical care over the past two decades for formerly underserved and deprived segments of the society cannot be all bad. Whatever our views in this regard, a mixed system (as Wildavsky acknowledges) is at present the only politically acceptable approach, given cultural and behavioral factors that inhibit adoption of either of the other two models.

The tendency to dichotomize between public and private (or governmental and market) solutions is misleading. It is not a question of either-or; it is a matter of drawing on both approaches to ameliorate the health cost problem (Kingsdale,

1978; Cohodes, 1982). The mixed system is a product of American pragmatism. It recognizes that people neither want to entrust their health needs to the whims of a purely private market nor depend for their satisfaction on a giant public bureaucracy. In this setting no dramatic improvements, only marginal changes, in the existing system are likely. The reform proposals with the greatest chance of adoption are those that fall within the framework of the mixed model, not those that seek to replace it. This is a lesson that the Reagan administration is now learning.[22]

NOTES

1. Health care spending has more than doubled since 1975 and is seven times greater than in 1965, when Congress created the Medicare and Medicaid programs.

2. As two health analysts observed in 1977, not too many years ago the number of political scientists interested in the politics of health care could have been counted on one hand (Friedman and Rakoff, 1977: xiii).

3. Even today, mainline political science continues to ignore the field, as indicated by the almost total absence of health-related articles in the standard political science journals.

4. The Medicare program consists of two parts: A (hospital insurance) and B (physician and surgeon fees insurance). The first is financed out of social security; the second, out of general tax revenue and contributory payments by covered individuals.

5. Although the United States is one of the few industrial countries of the world without a compulsory national health insurance program, over 85 percent of the population is covered at least in part by some form of medical insurance.

6. *San Antonio Independent School District* v. *Rodriguez*. U.S. Supreme Court Reports, 30 Lawyers Edition 2d. 16 (1973).

7. Members of the medical profession, for example, participate in their own regulation as agents of government through physician-dominated licensure and disciplinary boards and professional standards review organizations (PSROs). Similarly, government functions both as regulator and as purchaser of a major share of the health industry's output. This duality raises the likelihood that its interests as payer may influence its behavior as regulator (Pauly, 1981).

8. Federal subsidies to medical schools, for example, deprived the organized medical profession of control over the supply of physicians.

9. In Michigan, the "big three" auto manufacturers and their labor unions joined hands several years ago to push for cost-containment legislation. In St. Louis, twenty-eight of the largest corporations recently formed a health coalition to seek better controls over local medical costs. These examples are typical of action being taken by the private sector in many cities.

10. Third-party reimbursement, including Medicare and Medicaid, has made most hospital bonds a safe investment. The tax-exempt status that many of them enjoy (by virtue of being formally issued through a political subdivision) adds to their attractiveness.

11. A bill to limit the FTC's jurisdiction over licensed professionals is currently pending in Congress. Although strongly supported by the AMA, it has drawn the opposition of some segments of the health care establishment — including the Group Health Association of America (representing HMOs) and the American Nurses Association — which view it as an attempt to open the doors further to anticompetitive practices by the medical profession.

12. PSROs are panels of physicians who review the work of their colleagues to determine the appropriateness of Medicare and Medicaid admissions and length of stay in hospitals. In 1983, the Reagan administration proposed a new Professional Review Organization (PRO), which will replace PSROs. PROs will be statewide rather than local, may include non-physician groups (e.g., insurers) if physician groups do not apply or qualify, and will operate under two-year performance contracts. It is expected that half the existing 143 PSROs will be terminated as the 194 PSRO areas are consolidated into 52 statewide PROs (Urban Health, 1983: 14).

13. *Arizona v. Maricopa County Medical Society,* 50 U.S.L.W. 4687 (1982). Efforts by the FTC and the courts to eliminate unlawful restraint of competition in the health care industry are reviewed by Clark Havighurst (1980).

14. The HMOs, in effect, combine the function of provider and insurer.

15. A recent study by Lawrence D. Brown (1983) concludes that HMOs have failed to accomplish their fundamental objective — major reorganization of the health care system — because of flawed federal federal HMO legislation and the resistance of the politcal system to such basic change (see also Record, 1977, for the origins of the HMO concept).

16. The political unpopularity of making the consumer pay a greater share of the hospital bill was demonstrated in 1983 when the administration indicated that it was considering a means test for Medicare benefits under which elderly patients would pay a larger part of their medical costs. Public reaction was so severe that the head of the Department of Health and Human Services took the lead in vigorously denying that any such change was being contemplated.

17. At the end of 1981, only twelve states were in compliance.

18. Reagan officials contend that the CON process bars market entry to new providers and thereby inhibits competition.

19. The hospital industry is split over the question of retaining HSAs, with some groups, including the Catholic Hospital Association, looking to planning and CONs as a check on the growing number of for-profit hospitals. Such hospitals, they argue, constitute unfair competition by limiting their services and adopting restrictive admission policies that "skim off" the "paying" patients. (Space limitations have not permitted us to discuss the phenomenal growth in recent years of the investor-owned or for-profit hospitals.)

20. The administration has indicated its interest in a plan under which Medicare enrollees would be given vouchers to purchase their own health insurance. They would receive a bonus if they bought insurance costing less than the face value of the voucher, and they would have to pay extra if they wanted more expensive coverage. It is assumed that voucher recipients would purchase the most efficient plan, thereby encouraging competing insurers to offer less costly alternatives.

21. A persuasive case can be made that the vigorous promotion of proper dietary practices, safety laws, environmental pollution controls, and physical fitness programs would be more cost-effective in achieving better health than many of the proposed solutions to control costs. A growing number of corporations, for example, have established physical fitness or "wellness" programs for their employees with quite satisfactory results.

22. The Reagan administration may opt for a market-oriented health care system, but it is having second thoughts, if for no other reason than expediency, about its opposition to regulatory controls. These doubts have been instrumental in delaying the administration's long-promised plan for "overhauling" the health delivery system.

REFERENCES

Aday, L., R. Anderson, and G. Fleming (1980) *Health Care in the United States.* Beverly Hills, CA: Sage.

Alford, R. (1975) *Health Care Politics.* Chicago: University of Chicago Press.

Altshuler, A. (1965) *The City Planning Process.* Ithaca, NY: Cornell University Press.

Austin, C. (1975) *The Politics of National Health Insurance.* San Antonio, TX: Trinity University Press.

Brasfield, J. (1982) "Health planning reform: a proposal for the eighties." *Journal of Health Politics, Policy and Law* 6: 718-738.

Brown, J. (1978) *The Politics of Health Care.* Cambridge, MA: Ballinger.

Brown, L. (1983) *Politics and Health Care Organization: HMOs as Federal Policy.* Washington, DC: Brookings Institution.

Carey, C. (1974) "A constitutional right to health care: an unlikely development." *Catholic University Law Review* 23: 492-514.

Cohodes, D. (1982) "Where you stand depends on where you sit: musings on the regulation/competition dialogue." *Journal of Health Politics, Policy and Law* 7: 54-79.

Crawford, R. (1979) "Individual responsibility and health policies in the 1970s," pp. 247-268 in S. Reverby and D. Rosner (eds.) *Health Care in America.* Philadelphia: Temple University Press.

Dahl, R. (1964) *A Preface to Democratic Theory.* Chicago: University of Chicago Press.

Davis, E., M. Millman, et al. (1983) *Health Care for the Urban Poor: Lessons for Policy.* Totowa, NJ: Rowman & Allenheld.

Davis, K. and C. Schoen (1978) *Health and the War on Poverty.* Washington, DC: Brookings Institution.

Demkovich, L. (1983a) "Private sector moves in as Washington ends its financial assistance for HMOs." *National Journal* (September 3): 1787-1789.

Demkovich, L. (1983b) "Medicare on the critical list—can Congress plug a $200 billion gap?" *National Journal* (July 30): 1580-1585.

Enthoven, A. (1980) *Health Planning.* Reading, MA: Addison-Wesley.

Feder, J. (1977) *The Politics of Federal Hospital Insurance.* Lexington, MA: D.C. Heath.

Feldstein, P. (1979) *Health Care Economics.* New York: John Wiley.

Friedman, K. and S. Rakoff (1977) *Toward a New Health Policy.* Lexington, MA: D.C. Heath.

Greenstone, J. (1975) "Group theories," pp. 125-165 in F. Greenstein and N. Polsby (eds.) *The Handbook of Political Science,* Vol. 2. Reading, MA: Addison-Wesley.

Harrison, D. and J. Kimberly (1982) "Private and public initiatives in health maintenance organizations." *Journal of Health Politics, Policy and Law* 7: 80-95.

Havighurst, C. (1980) "Antitrust enforcement in the medical services industry." *Milbank Memorial Fund Quarterly* 58: 89-124.

Hayes, M. (1978) "The semi-sovereign pressure groups." *Journal of Politics* 40: 134-161.

Hyman, H. (1973) *The Politics of Health Care.* New York: Praeger.

Joskow, P. (1981) "Alternative regulatory mechanisms for controlling hospital costs," pp. 219-257 in M. Olson (ed.) *A New Approach to the Economics of Health Care.* Washington, DC: American Enterprise Institute for Public Policy Research.

Kallstrom, W. (1978) "Health care cost control by third party payors." *Duke Law Journal* 6: 645-697.

Kingsdale, J. (1978) "Marrying regulatory and competitive approaches to health care cost containment." *Journal of Health Politics, Policy and Law* 3: 20-42.

Leichter, H. (1979) *A Comparative Approach to Policy Analysis: Health Care Policy in Four Nations.* Cambridge: Cambridge University Press.

Lippincott, R. and J. Begun (1982) "Competition in the health sector: a historical perspective." *Journal of Health Politics, Policy and Law* 7: 460-487.

Marmor, T. (1973) *The Politics of Medicare.* New York: Aldine.

Marmor, T. and J. Marone (1980) "Representing consumer interests: imbalanced markets, health planning, and the HSAs." *Milbank Memorial Fund Quarterly* 58: 125-165.

Nathan, R., F. Doolittle, et al. (1983) *The Consequences of Cuts: The Effects of the Reagan Domestic Program on State and Local Governments.* Princeton, NJ: Princeton Urban and Regional Research Center, Woodrow Wilson School of Public and International Affairs, Princeton University.

Pauly, M. (1981) "Paying the piper and calling the tune: the relationship between public financing and public regulation of health care," pp. 67-86 in M. Olson (ed.) *A New Approach to the Economics of Health Care.* Washington, DC: American Enterprise Institute for Public Policy Research.

Poen, M. (1979) *Harry S Truman versus the Medical Lobby.* Columbia, MO: University of Missouri Press.

Record, J. (1977) "Medical politics and medical prices," pp. 71-106 in K. Friedman and S. Rakoff (eds.) *Toward a New Health Policy.* Lexington, MA: D. C. Heath.

Rosen, G. (1958) *A History of Public Health.* New York: MD Publications.

Savas, E. (1982) *Privatizing the Public Sector.* Chatham, NJ: Chatham House.

Steinwald, B. and F. Sloan (1981) "Regulatory approaches to hospital cost containment," pp. 273-308 in M. Olson (ed.) *A New Approach to the Economics of Health Care.* Washington, DC: American Enterprise Institute for Public Policy Research.

Thompson, F. (1980) *Health Policy and the Bureaucracy: Politics and Implementation.* Cambridge, MA: MIT Press.

Thurow, L. (1980) *The Zero Sum Society.* New York: Basic Books.

Urban Health (1983) "Watching Washington (monthly column)." July: 14-16.

Weller, G. (1977) "From pressure group politics to medical-industrial complex." *Journal of Health Politics, Policy and Law* 1: 444-470.

Wildavsky, A. (1979) *Speaking Truth to Power: The Art and Craft of Policy Analysis.* Boston: Little, Brown.

8

Politics as Accusation:
New York's Public and
Voluntary Hospitals

FRED H. GOLDNER

□ WE ARE ALL FAMILIAR WITH the claims that politics
is filled with rhetoric and is heavily dependent on symbols.
Similarly, many scholars of organizations have long used a
political model to understand organizational processes. Politics, however, is also a label that is used by organizational
participants in a system to describe actions of which they
disapprove. The word "political" is often used as a pejorative
adjective to discredit an organizational action or even to
characterize an entire organization. These uses have been
employed especially in disputes about the relative character of
public versus private organizations and, as is the focus of this
chapter, specifically in disputes about health care organizations
in New York City.

I will attempt below to highlight the uses to which this
appellation of politics is put, the degree of its accuracy, and its
utility as an ideological weapon that can both reveal and mask
the degree to which public and private nonprofit hospitals and
agencies differ from each other. I will do this by recounting
some of the reactions of the public and private hospitals of New

AUTHOR'S NOTE: *This material is based on work supported by the National
Science Foundation under Grant SES 800893 and the Yale University Program on
Non-Profit Organizations. Any opinions, findings, and conclusions expressed in this
chapter are those of the author and do not necessarily reflect the views of these
sponsors. I also thank the Faculty in Residence Program of Queens College.*

York to an attempt to regionalize their services; by the story of how one "establishment" institution's quiet diplomacy no longer served it in the face of a rejection of its plans to build a burn center; by an account of competition among ambulance crews for reimbursable patients; by illustrating the rancor that characterized opponents who were both symbiotic and antagonistic; and by providing some examples of political hirings and firings.

POLITICS AS AN ACCUSATION

My introduction to the charge that political considerations would play a large and negative role in any health organization whose management was not totally divorced from the public sector came in October 1975 in hearings held by the City Club of New York. These hearings functioned to present arguments about the choice of groups that should receive federal designation and monies to create and run a health systems agency (HSA) for New York City. Congress had provided for the creation of such HSAs throughout the country to plan health care for their areas and to review requests for capital and major equipment expenditures from hospitals receiving federal reimbursement monies from Medicare and Medicaid. The HSA for each area was to replace the different planning agencies that had previously been set up to oversee in part the various federal programs that had been created throughout the earlier years to provide health care monies. The two main contenders for designation were a group directly out of the city administration's Comprehensive Health Planning Agency (CHPA) and a group out of the United Hospital Fund (representing the voluntary — or private, nonprofit — hospitals in the city) and the Health and Hospital Planning Council of Southern New York. The CHPA had been in existence only about five years. The Health and Hospital Planning Council and its immediate predecessor organizations had been controlled by representatives of the voluntary hospitals for over thirty years as the planning group for the major private providers of hospital care and had had quasi-legal responsibility for review of all proposed new hospital facilities since 1965.

Speaker after speaker claimed that providing the designation to those who ran the CHPA would lead to harmful effects, because the resultant organization would be too political, whereas the attack on the establishment group was made on the basis of their lack of accountability. As a union leader testified, "Experience has shown that profit-making and voluntary non-profit health agencies are willing and able to absorb public money in their operation but are unwilling to submit to public scrutiny and accountability in their operations and the expenditure of their funds."[1] It is significant to note that the "private" group never bothered to answer these charges of accountability. On the other hand, CHPA supporters found it necessary to answer the political charge. They did so, however, not by denying it but by asserting that success would depend on access to and hence some connection to elected public officials whose cooperation they claimed would be needed for effective planning. The one thing both sides did hold in common was their distrust of each other.

The remarks of Monsignor James Cassidy, Director of Health and Hospitals and Public Charities of the Archdiocese of New York, were typical of those who raised the political issue at the hearings. After speaking against the CHPA and the people in it, he made it clear that he wanted a voluntary agency that was not controlled by the city. In giving his reasons he said, "we would hate to see that the health of the people of the city of New York be hindered by the political pressure or the individual needs of particular people, or power plays involved. We want an agency that's really going to serve the needs of our people." I have gone to the trouble of quoting Monsignor Cassidy because the contrast between that statement and subsequent remarks made by him call into question whether the political accusation was his actual assessment of the facts or whether it was merely a rhetorical device used to defeat one group in favor of another with which he was more closely identified. The remarks I cite below were made one year after those already quoted and resulted in the political epithet now being hurled at him.

First, however, I must identify the institutions referred to below. HHC refers to the Health and Hospitals Corporation, a semiautonomous organization that ran the seventeen municipal hospitals of New York City, representing 20 percent of the beds and 50 percent of the clinic and emergency room visits in the city. Monsignor Cassidy and others in the City Club hearings frequently cited HHC as a negative example of a politically run system — in contrast to the voluntary hospitals. A "voluntary hospital" is the term used to describe the common general-purpose, acute care hospital with which most people are familiar. They are private, nonprofit organizations, in further contrast to profit-seeking hospitals, which are referred to as proprietary hospitals and of which there are relatively few in New York. HHC contracted with a number of voluntary hospitals and medical schools for the provision of professional services at most of its seventeen hospitals. Misercordia was a voluntary hospital identified with the archdiocese represented by Monsignor Cassidy. Albert Einstein Medical College was identified with the Federation of Jewish Philanthropies.

The New York Times on October 20, 1976 (p. 1), quoted Monsignor Cassidy on whether the HHC should sign an affiliation contract for the provision of professional services for its new Lincoln Hospital with either Misercordia Hospital or with Albert Einstein Medical School. "Catholics," he stated, "are tired of being pushed to the wall on this kind of thing." The *Times* then reported that Monsignor Cassidy "said that Jewish-supported hospitals and medical colleges had a big share of the city's lucrative medical affiliation contracts at municipal hospitals while the Catholics had none. . . . Monsignor Cassidy warned that if Misercordia was denied a major affiliation contract to replace the minor one it now had at Lincoln, the Beame administration would risk the opposition of Catholic voters in the Bronx" (p. 16). I do not know if anyone would seriously have tried to make the claim that Misercordia was better able to provide a quality affiliation or that patients would receive better care under Misercordia than under Albert Einstein, but Monsignor Cassidy made no mention of even such a claim. There was a reaction from Albert Einstein Medical School on the same page. Its dean was quoted in the article to say that any move to oust his institution was a "blatant political act."

Misercordia Hospital's affiliation with Lincoln Hospital was imposed on the HHC by city officials. Largely because their proposal more nearly met federal specifications, the officials from the CHPA did get the designation as the group to form the HSA — but it would be as a formally private association, albeit with some city officials on its board, including the provision that the city health commissioner would serve as chairman of its executive board.

DIFFERENT STYLES OF POLITICS

Between these two events I went to work for the Health Systems Agency, serving for eight months in the capacity of deputy director and followed this by sixteen months at the Health and Hospitals Corporation as chief-of-staff and executive deputy to the president.

From the HSA I witnessed the negative views that participants in the voluntary and municipal sectors held of each other as well as their different styles of political activity. These views and styles were evident from the reaction to HSA's regionalization plan, which was drawn up in answer to a request by the mayor to look at and make recommendations on the entire system of hospital care in the city. The plan was focused on the need to reduce the costs of the hospital system, since the city paid 25 percent of all Medicaid costs as well as a subsidy to the HHC. The key ingredients of the plan were attempts to provide consolidation and specialization in order to reduce costly duplication produced by each hospital's desire to provide a full range of services. Although the plan recognized the unique role played by HHC, it differed from most of the thirty other lengthy studies of the hospital situation in New York City produced since 1950 in that it did *not* single out HHC as the only cause of problems in hospital care, and it differed from all of them by proposing a set of rewards and sanctions by which the state could induce medical institutions to accept the plan.

I was immediately struck by the apparent defensiveness of those affiliated with HHC and the intensity of their feelings about the voluntary sector. For example, when the plan was presented to a meeting of the HHC board of directors, they reacted by taking a number of verbal swipes at the voluntary sector, with one member asking sarcastically, "Is this acceptable to the powerful sector — the voluntaries?"[2] When we met with the Council of Medical Boards of the various HHC hospitals, we found that although many were employed by the voluntary hospitals or medical schools that provided the professional staff of the municipal hospitals under the affiliation contracts, they apparently identified with the municipal system. They saw the voluntary hospitals as having greater political power and resented it: "Municipals are still being attacked and not the voluntaries," "Voluntary beds are still being built," "Bad voluntaries are staying open," "Some institutions have the ability to resist and others do not."

The presentation of the regionalization plan to a joint meeting of the various community boards of the HHC hospitals met with a negative reaction which echoed that of HHC's board of directors and the medical boards — not enough for us and too much for the voluntaries. One member put it, "Why not a more powerful and better funded public system?" Attacking the private nature of the voluntaries, it was stated, "Nowhere in the plan did I find anything on measures of public accountability of money and care." Others saw the plan as helping the voluntaries at the expense of HHC: "I find when we look at your organization that we are swallowed. We find ourselves trying not to be swallowed by voluntaries and big powerful trustees. Even Mayor Beame must take into account the big power." Still another put it more succinctly when referring to the report: "It's called the voluntarization of the municipals." One response to the plan incorporated the accusation of politics in comparing the two systems: "The municipal system has serious political problems that voluntaries don't have. How does the plan make them equal? The municipal system needs protection. Will HSA protect the municipal system?"

On the other hand, a representative of one of the voluntary hospitals reacted to the regionalization plan as an action against the voluntaries: "We're a little paranoid. The plan was seen as a bailing out of the corporation by spreading their deficit around all the hospitals in the city." His reaction, however, was not typical. At that early date in HSA's history little was heard from the voluntary hospitals. They appeared to be rather unconcerned with any systemic plans to cut medical costs by sharing facilities or by giving up any of their specialities, regardless of costliness. Most appeared to resent the notion that an agency such as HSA had any role to play, and they essentially ignored the health systems agency until the burn center decision" (referred to below).

It appeared that the voluntaries assumed they would be able to continue to affect any governmental policy by continuing their style of politics, which was more akin to quiet diplomacy among their leaders, key members of their boards, and the leading politicians — that is, among a few gentlemen who were comfortable with their mutual familiarity over the years. Such political strength was referred to in notes on an early draft of the regionalization plan that HSA had received from staff members of the state-created unit that was set up during New York City's fiscal crisis to monitor the city's budget — the Emergency Financial Control Board: "There is little recognition in the plan of the very real barriers to implementation (e.g., political ties of voluntaries . . .)."

HHC officials employed the opposite of quiet politics. For example, almost concurrently with the regionalization plan, HSA served as staff to a special mayoral commission set up to investigate the internal workings of HHC. It appeared from that vantage that HHC was mismanaged, as the commission concluded in its report. HHC officials, both during and after the investigation, complained long and often about injustices to the poor. Here, they appeared to think that wrapping themselves in the cloak of protecting the downtrodden would make up for their lack of attention to the management of their resources.

But for both sectors it was politics as usual — although the usual was different for each. One attempted to continue, through quiet diplomacy among a few power brokers of like minds, a form of establishment politics. The other believed that their felt injustices were so obvious that by screaming about them all their other troubles would end — a form of public or mass mobilization politics. Both these approaches appeared to be anchored in the past.

It is important to note here that the voluntaries were not a coherent or organized group. On the contrary, they largely played their politics separately, the more prestigious medical centers having little or nothing to do with any other hospital. Moreover, there were great differences in the power of even the weaker hospitals, as can be seen from the easy demise of some, the partial buyout of others, and the rescue by extraordinary means of still others, such as Brooklyn Jewish Hospital, which *The New York Times* has described as "politically well-connected" (March 8, 1979, p. B5).

POLITICS AND THE
REJECTION OF POLITICS

The HSA did engage in political behavior and would have had to do so, regardless of the auspice of its origins, if it were to accomplish anything, for its large board (over ninety members) was both constituency and geographically based, and a federal mandate required a majority of consumer and a minority of provider representatives. This made it difficult for such a group to arrive at a consensus about anything important that was the least bit controversial. Consequently, some of its business, such as the review process, was subject to the approval of a smaller executive committee only. Moreover, the initiative for action lay in the hands of the full-time staff, who then had to be politically adroit in the additional task of dealing with, influencing, and soothing individual board members.

The HSA's planning function, unlike its review function, was without any enforcement powers of its own. The staff, however, succeeded in raising issues and setting agendas by doing such studies as the regionalization report. Although these reports required the approval of the full board, the staff was able to effectively bypass that procedure or render it ineffective. For example, it did not matter that the regionalization report was never officially passed by the board. It had become a public document the minute it was distributed to that large board, and the newspapers, as they usually do, treated the draft as if it were a final document. The mayor, in turn, therefore, was free either to act on it or to hold off by claiming it was not an official document.

Perhaps of more interest is that while that plan was produced in response to a request from the mayor, it originated in talks between HSA staff and city officials. The mayor's letter of request was actually drafted by those to whom it was addressed. The staff used such political means to attempt to achieve changes in the system because there simply was no other institutional base for the origination of such initiatives.

The complacency of the voluntary hospital sector was shattered on an occasion when the HSA successfully resisted severe political pressures that had been mobilized by the most prestigious voluntary hospital in the city. The HSA, acting in its other capacity of reviewing capital requests, turned down an application from New York Hospital to build a burn center that was to be housed in a new hospital for plastic and reconstructive surgery, after the staff produced evidence indicating a number of weaknesses and errors in the maze of financial data and market assumptions used by the accounting firm in the supporting material submitted by the hospital. The applicant had secured the support of the mayor and deputy mayor who put considerable pressure on the HSA staff to support the project. The applicant also had the enthusiastic support of the fire commissioner, who encouraged firemen to attend the final hearings for the application. In fact, the audience was so large that the hearings had to be moved from the usual HSA hearing room to a rented university auditorium.

It would have been easy to approve the application, for the weaknesses and errors, which were not readily apparent, were exposed only after great time and effort. Without both the acknowledgment of the need for some politics and the resolve to be fully conscious of the choices that could be made among

the various pressures, the time and effort to produce the necessary data would never have been spent. Thus, I am not here
trying to make the naive assertion that the HSA staff believed
that only technical considerations should or could be employed
in their work. On the contrary, they were keenly aware of the
necessity of taking political considerations into account and of
using political methods. No organization operating in such a
complex environment could succeed without being somewhat
of a political creature. The problem is one of finding a balance
between political acts and technical or merit issues that directly
bear on the goals of the institution. In this instance, the staff
director resisted the orders of the deputy mayor and the threats
of his assistant on the basis that "it was as good an area to fight
as any," given the data he then had.

Rather than being more political than an HSA created from
the former "private" Health and Hospitals Planning Council,
this one was probably less so on this first key review decision.
A result was that the prestigious voluntary hospitals could no
longer assume the effectiveness of quiet diplomacy. Their network of influence did not include those responsible for this
decision. As one city official commented after the HSA action,
"Luckey and Thompson [the heads of the New York Hospital
and Medical Center] will call up the president of Chemical
Bank or First National and none of them will know what to do.
Their [hospital] is in traditional circles."

The accusation of "politics" aimed at the CHPA group by
its competitors should now be seen as a political weapon itself.
Any HSA would have had to employ politics to some degree.
As we have seen, private voluntary hospitals attempted to
mobilize political support whenever they could. Those HSA
competitors had themselves been political creatures, albeit
through the quiet politics of private influence. That point
(though largely ignored) was made by one speaker during the
City Club hearings: "Assertions that technical merit alone
governs the decisions of voluntary agency planning, that politics played but a trivial role in these decisions, can provoke but
derisive laughter from the *cogniscenti* with intimate knowledge
of how the system really works. . . . I attribute no human
villany to the political process. It is understandable and predictable that the representatives of the voluntary hospitals have

been loath to reject each other's expansionist plans." The more traditional review committee would have been less likely to call into question the data submitted by New York Hospital that had been prepared with the expertise of a major accounting firm. You do not do that to friends. You confront them only if the data are easily and publically available.

PUBLIC AND PRIVATE NECESSITIES
FOR POLITICS

As I mentioned earlier, HHC had been used by CHPA opponents as a negative model of a politicized public organization. Even the author of the long quotation that characterized the voluntary planning agencies as political disavowed HHC by stating, "Heaven forbid that I would support any replication of that kind of a public creature." He was careful to distinguish any HSA from HHC on the basis that HHC was dependent on city funds, whereas the HSA would not be dependent on local politicians because it was to be federally funded. These criticisms of HHC carried with them the obvious, though implicit, comparison to voluntary hospitals. I was able to evaluate some of those implicit claims about the difference between HHC and voluntary hospitals from within HHC, for after the eight months with HSA I then spent from May 1977 to August 1978 as the executive deputy to the president of HHC.

In one important respect, neither HHC nor the voluntaries could avoid being political. Although HHC received subsidiary monies directly from the city, it received the bulk of its monies through third-party governmental reimbursement schemes such as Medicaid and Medicare. The voluntaries also received much of their monies from those same sources, and almost all of the rest from state-regulated third parties such as Blue Cross, rather than directly from patients. These reimbursement systems were complex ones based on numerous interpretations and definitions that permitted or required judgment by governmental administrators. Trying to influence administrative judgments was, of necessity, a political process for both the voluntary and the municipal system, requiring the

development of a network of contacts with financial people
from state, city, and even federal agencies. The hospitals and
those administrators needed each other to accomplish their
tasks. The development of such friendships, or even personal
familiarity, enabled them to share information and to try to
provide the kind of elaboration about their problems that would
influence judgments.

Another, less subtle form of politics was involved in the
competition that existed between the municipal and voluntary
hospitals. The reimbursement system for all hospitals was
structured in such a way that there were heavy penalties for
empty beds, especially if a hospital's patient load was declining
or fell below a specified occupancy rate. This resulted, therefore, in a good deal of competition for "paying" or "covered"
patients among the hospitals.

HHC did not compete very effectively, however. It was
obligated to accept patients regardless of their ability to pay,
and it did not have enough patients who were covered by any of
the various forms of insurance to make up for the large number
who were without such coverage. These medically indigent
patients were neither poor enough to be eligible for welfare and
Medicaid nor wealthy enough to afford to pay hospital costs.

When we entered HHC, we announced a number of programs that would enable the corporation to compete more
successfully for patients by upgrading hospital services and
making our facilities more attractive. During the first month in
office we met with representatives of the voluntary hospitals
and medical schools with whom HHC had affiliation
agreements to inform them of our plans and to tell them that the
payments for their services would have to reflect the realities of
a decreasing or increasing patient census. They evidently
thought that the municipal hospitals should confine themselves
to those patients who were "uncovered" or too poor to pay for
care, because they saw the announcement as some kind of new
threat. As one of them put it with an implied threat of his own,
"I always thought the competition between the voluntaries and
municipals was just talk, but if you are going to aggressively
recruit patients. . . ."

The one area where the voluntary hospitals had become
more aggressive in recent years had been that of emergency

room visits. During the period 1970-78, the percentage of clinic visits to the 17 municipal hospitals had increased from 49 to 52 percent, with a correspondingly decreasing percentage going to the 66-odd voluntary hospitals. At the same time, however, the emergency room visits to municipals had declined from 52 to 46 percent. The reason for the corresponding increase of emergency room visits to voluntary hospitals and their decreasing percentage of clinic visits was clear. Everyone lost money on clinic visits because the state had more stringent reimbursement maximums on outpatient care than on inpatient care. Emergency room visits, on the other hand, were more likely to lead to inhospital stays, particularly if patients were brought there by ambulance as a result of trauma. The hospitals were aware of these differences and had been acting accordingly, especially those who ran their own ambulances.

THE POLITICS OF AMBULANCES

One of the voluntary hospitals politicized the competition when they publicly fought against the cancellation of a contract HHC had had with them that subsidized part of their ambulance costs. Again, ambulances brought people to emergency rooms who were then likely to end up occupying reimbursed beds. The Emergency Medical Service division of HHC provided the bulk of ambulance services to the city under the 911 emergency phone number shared with the police and fire departments. In some areas, however, EMS contracted with voluntary hospitals for them to cover specific geographic areas and paid them accordingly. When we came into office in 1977, we determined that the Emergency Medical Service needed to be rapidly upgraded. Accordingly, we created a vice-presidential-level position to run EMS and filled the position with a former high police official. After substantially improving the service, he came in at the end of November 1977 with an analysis that called into question some of the subsidies. One of them stood out. We were paying Cabrini Hospital $166,000 to operate two ambulances in their area, although their hospital was within seven blocks of one of the major trauma centers in

the country — HHC's Bellevue Hospital. Cabrini was informed that their contract would not be renewed on March 1,
1978.

The roof caved in. With the backing of the Greater New
York Hospital Association, the hospital fought a political war
and almost succeeded in removing the entire ambulance service
from the public sector. They yelled because they too needed to
fill their beds and they did that with a large number of patients
who came through the emergency room (in contrast to the more
prestigious voluntary hospitals, where almost all patients are
admitted through scheduled admissions by private physicians).
They took out newspaper advertisements, mobilized a constituency connected to local politicians, and besieged the mayor's
office. The *New York Daily News* (February 27, 1978, p. 4)
accurately characterized the political nature of the situation:
"The war is fought on the streets, where ambulance drivers
from rival hospitals race for bodies. On another level, the
political battle for control of 911 is fought in hospital and health
care council and board rooms and has reached into the mayor's
office."

The advertisement they placed in *The New York Times*
(February 22, 1978, p. A20) claimed that the "cancellation of
the Cabrini contract is the first step in an acknowledged plan of
the New York City Health and Hospitals Corporation to take
over emergency ambulance services in the city. Full implementation of this plan will mean that Voluntary Hospitals no longer
provide emergency ambulance service. City ambulances will
pick up all emergency care patients. And they will transport
them to their base hospitals — the city hospitals." None of this
was true. We had offered to let their ambulances remain on the
911 system but without our subsidy. They ignored that offer in
their ads in the obvious attempt to enlist the aid of all voluntaries who had become uneasy about our aggressive management. In one phone conversation their executive director
insisted the subsidy be continued on the grounds that to do
otherwise would jeopardize the subsidy paid to other voluntary
hospitals, despite the obvious uniqueness of duplicative service in their case.

The major charges that Cabrini raised against HHC in the
evident attempt to enlist public and political support was that

our ambulances had been raiding their territory by taking patients to Bellevue instead of to Cabrini. As the newspapers put their charge: HHC had been engaging in "body snatching" by opting to fill beds at municipal hospitals even if they were located further away than a voluntary facility such as Cabrini. Those were fascinating charges, because our data indicated that the accuser was more sinner than sinned against. We had our staff take a two-month period that coincided with our arrival at HHC in the spring of 1977 to sustain or refute their charges. During that eight-week period, Bellevue-based ambulances had made 174 calls out of their assigned area in which they brought patients back to Bellevue — tending to confirm the accusations. However, in that same period, the three neighboring voluntary hospitals had done the same thing on 665 such calls, and 227 of them were made by Cabrini ambulances. We then took current data from January 10 through 31, 1978, to see if things had changed; they had not. In that period Bellevue made 34 such calls to the 133 such calls by the same three voluntary hospitals, an even higher ratio in Bellevue's disfavor than before. In fact, in that later period those kinds of ambulance runs accounted for 30 percent of all Cabrini's runs, a higher percentage than that of any of the other hospitals.

The voluntaries and Cabrini in particular were now as liable as anyone to the charge of "politics," and the union representing public hospital workers (a different one represented voluntary hospital staffs) took out their own advertisement in *The New York Times* (March 30, 1978, p. C20) to do so. After pointing out an unchallenged assertion about Bellevue's superior service, they claimed that the ambulance controversy was "not being discussed responsibly" and that "political and special interest groups competing for the service have yet to demonstrate any proof of capability to run the Emergency Medical Service." They concluded with a plea for the elimination of political criteria:

> Now is the time to watch the service, study the service, away from the headlines, and to develop improvements in the best, objective, nonpolitical manner.

Our decision to cancel the Cabrini contract turned out to be a mistake. Cabrini's claim in its advertisemént quoted above, that the cancellation was "irrational," turned out to be true if one includes in the concept or rational the facts of political power. We had been right in every respect except that we had misjudged the political factor in this case. We had not accurately gauged the degree of latent hostility or fear in the voluntary sector or the ability of Cabrini to mobilize political support.

PERSONNEL POLITICS

I have tried to illustrate that private, nonprofit organizations may be involved in political action as much as public ones and may even be able to utilize political tactics more successfully than their governmental counterparts. But somehow the accusation of organizations as dominated by politics continues to flow in one direction; in fact, in many of the places where the accusation was true, such as the awarding of affiliation contracts, it was because those who were hurling the accusation were those who were responsible for the politicization.

I do not maintain that there were no areas where the political characterization was truer of the public-sector hospitals than the private. The one area where it certainly held was in the appointment and termination of the heads of the HHC. I have made the point elsewhere (Goldner, 1983) that the public-sector organization is under a comparative disadvantage because of the nature of the political turnover of elected officials. Mayors of New York have looked upon those who headed municipal organizations such as the hospital corporation as part of their staff rather than as executives who had to run a complex organization. Although the president of HHC nominally serves at the pleasure of the board, the mayor has had de facto power to name or remove presidents ever since the removal of the president in early 1977. All of the six succeeding presidents or full-time board chairmen who served under Mayor Koch between 1978 and 1982 were appointed and/or removed by him.

The president under whom I served was suspect from the beginning of Mayor Koch's term merely because he had been appointed by Koch's predecessor. Elected officials frequently continue to run against past administrations, and Koch was no exception. This is seldom true in private organizations, non-profit or otherwise, because their leadership is largely self-perpetuating and not subject to public scrutiny.

During my tenure at the corporation, politics played no direct role in the composition of the senior corporate staff except for the general atmosphere of racial politics. We understood the necessity of having minority representation in those appointments. It made good organizational sense because our patients were largely of minority status, as were many of the hospital staffs. It simply would not have been right or proper to have a totally nonminority senior staff.

One senior appointment was made with another constituency in mind. We hired someone from the private, profit-making sector for the position of vice-president for finance. There seemed to have been an assumption on the part of those who played a role in city finances, such as Emergency Financial Control board members and bankers crucial to the city's future, that they were more likely to relax if we obtained the kind of financial expertise that they saw as available only from the profit-making sector. As it turned out, however, they were wrong. There is a vast difference in what it takes to succeed in the municipal sector under constant public scrutiny from what it takes to operate in the private sector without such scrutiny.

The one position that was more frequently involved in politics was that of the executive director of the hospitals. Here it was racial or, perhaps more accurately, community politics. Hospitals that were in communities that were clearly dominated by this or that ethnic group played a major role in those communities, because they frequently were the only major institutions in those neighborhoods as well as one of the largest employers. Therefore, as an example, it was clear that we had to find an Hispanic director for Lincoln Hospital, which was an integral part of the Hispanic South Bronx.

My own leaving of the corporation was political but in the ordinary way that politics plays a role in all organizations — public or private. A recently appointed full-time board chairman attempted to seize all power from the president in an office coup. He met with the president one Sunday night and handed him a list of demands headed by my leaving the next day and followed by the president's moving into my office (thereby enabling the chairman to have two large offices) and the vice-presidents reporting directly to him instead of to the president. I was scheduled to leave to return to the university in a few weeks' time, so there was little to be gained by fighting over that. The president essentially won the other battles, and the chairman followed me out by a few weeks, when he was forced to resign over a number of issues that had become public through leaks to the television and press media. This, I maintain, was no different from many such power struggles that take place within all organizations among executives, either on personal bases or on differences of policy. The chairman knew that many of us opposed much that he wanted to do and not do, and he took after the most obvious and vulnerable target. Let me return for a moment to the burn center controversy to quote one of the officials at New York Hospital as he discussed the internal opposition from other specialities over the potential growth of their burn unit: "Politicians would be green with envy to watch the way our internal politics work."

To reiterate, my leaving was political in the ordinary way that politics plays a role in all organizations. But after I left I was at the receiving end of the political accusation. This example will serve to illustrate the pejorative use of the term "politics" and its careless identity, outside any kind of reality, as a property of public organizations. After the chairman left, I went back to the offices a number of times, leading to a rumor that I was returning and resulting in a planted story in the newspaper that I was coming back as a consultant (which indeed was being discussed) with, as the story had it, an "office, an executive secretary and a chauffeur-driven city car" — all of which was manufactured (*New York Daily Press*, September 21, 1978, p. 5).[3] Despite denials, the paper accepted it as fact and in an editorial the next day labeled it as a political act:

> The quasi-independent agency that runs the city's 17 munici-
> pal hospitals was created with the aim of removing an im-
> portant government body from the hands of politicians. Yet to
> our knowledge this has not happened. In fact, it appears that
> the political grip has tightened. For example, a Queens Col-
> lege professor is being paid a consultant fee of $190 a day and
> given his own car and secretary, while still drawing $25,000 a
> year from the college [*New York Daily Press*, September 23,
> 1978, p. 19].

It remains totally unclear to me how I represented a "political grip." I had no constituency whatsoever or the backing of any politician, only the respect of the remaining executives of the corporation.

CONCLUSION

I have tried to illustrate how organizational behavior incor-
porates the use of pejorative words as ideological weapons.
More specifically, I have attempted to show that the term
"politics" has been used as such a weapon in the relationship
between public and private sectors of the hospital industry,
thus revealing the ubiquity of the political process itself.

I have not attempted the still necessary task of making
distinctions among the many uses of the political appellation. It
clearly means, at one time or another: the actions of elected
officials, their interference in private organizations, and the
reverse; the struggles over the allocation of scarce resources
among the many claimants within each organization; the han-
dling of personnel on other than obviously objective and
agreed-upon criteria; and individuals who attempt to advance
their careers through interpersonal manipulation where those
skills are not part of the job, rather than by merit.

I hope I have succeeded in showing how one group has been able to take advantage of a general belief about the exclusive identity of political acts with the public sector both to disparage that sector and to mask its own, similar involvement. This identification of politics with any public-sector organization perpetuates a distinction that increasingly ceases to exist.

Merely to point that out, as this chapter has tried to do, becomes a kind of political act in itself, because it automatically becomes a defense of the public sector to do so. No relationship in our society is filled with more rhetoric than that between public and private organizations. Scholars ought to devote more energy to analyzing and clarifying the differences between and similarities of these two sectors, as well as the roles played by belief systems about those differences.

NOTES

1. The quotations from these hearings come from transcripts of tapes I made as the hearings were broadcast over radio station WNYC

2. Much of the material in this chapter comes from notes I kept during and after my stay in office.

3. The *New York Daily Press* was a newspaper created during the newspaper strike in New York City in 1978. The reporter of this piece was the regular hospital reporter for the *New York Daily News*.

REFERENCE

Goldner, F. H. (1983) "The daily apple: medicine and media in New York," in V. Boggs et al. (eds.) *The Apple Sliced*. South Hadley, MA: Bergin.

9

Urban Health Care:
Change and More Change

ELI GINZBERG

☐ AMERICANS STILL HAVE AN UPBEAT philosophy of life. Despite recurrent wars, the Vietnam debacle, Watergate, nuclear instability, inflation, unemployment, and many other evils, we believe, although probably less strongly than in the past, that if only we make the effort, we can reduce if not eliminate much that is wrong, inequitable, and inefficient in both the private and public sectors of our society.

Nowhere has this upbeat conviction been more pronounced than in the case of health care, particularly since the days of the Great Society ushered in an avalanche of new federal initiatives to improve care — principally the care of the elderly and the poor. Since seven of every ten persons to benefit have resided in standard metropolitan statistical areas, the thrust of the new initiatives has had heightened importance for urban health.

A single session of Congress in the mid-1960s passed no fewer than forty pieces of health legislation that the most activist of all U.S. presidents, Lyndon Baines Johnson, eagerly signed. Before we assess developments in the period between 1965 and today and peer into the future, it is desirable to recall what the urban health scene looked like at the time of the big push forward.

The first point worth recalling is that whatever category of city we focus on — the largest metropolitan centers such as New York, Chicago, and Los Angeles; the next in size, such as Cleveland, Dallas, Denver, and Boston; or those in the third rung, such as Nashville, Tucson, and Columbus — considerable variation has existed with respect to the quantity and quality of health care services available in these different communities and particularly the ease of access of the elderly, poor, and near-poor to physicians, hospitals, clinics, and other providers.

The important critical differentiators with respect to access to health care included the following:

— The strength of the medical infrastructure, particularly the number of medical schools, teaching hospitals, and specialists. On this score, New York, Boston, and Nashville had the edge — often a marked edge over the other cities in their class.

— The strength of the philanthropic tradition that helped to provide ambulatory and inpatient care to persons who were unable to pay or who could not pay all of their bills.

— The scale of public expenditures — state, county, local — for and on behalf of the poor and the near-poor either as direct providers of care or through reimbursement of other providers.

— The proportion that racial and ethnic minorities formed of the total population and established practices for dealing with these groups — whether through segregated or nonsegregated facilities.

A schematic assessment based on the foregoing suggests that the following was a reasonable approximation to the nature of urban health care on the eve of the major reforms in 1965:

— The employed middle class, through income and insurance, was able to obtain most of the medical care it needed, running into difficulties usually only in the event of catastrophic illness.

— In cities such as New York and Boston, where teaching and municipal hospitals have long cared for large numbers of the medically indigent, many poor or near-poor were able to obtain routine and emergency medical care. If they were admitted to a major voluntary hospital, the care was likely to be good. Most public hospitals were under severe pressures in staffing. Many were forced to accept mostly foreign medical graduates into their residency programs.

— The relocation of many middle-class whites from inner-city locations to the suburbs had served as a magnet to draw out of these neighborhoods many private practitioners who had earlier practiced in the inner city and, what is more, discouraged new graduates from opening practices in these inner-city areas. As a consequence, more and more of the urban poor had to obtain their primary care from the emergency rooms or outpatient departments of nearby hospitals.

— The elderly who were no longer employed and who were therefore without satisfactory or any insurance coverage were at serious risk if they required hospital care. Many were forced to use up all of their savings to cover a serious episode and seek financial help from their children to meet their bills.

— The slow expansion of U.S. medical schools in the 1950s and early 1960s in the face of a much enlarged demand on the part of consumers for more ambulatory and inpatient care resulted in a tight market for physician services in which many patients had to wait several days and often several weeks for an appointment. That more and more physicians were limiting their practices to a single specialty and were restricting their office hours further broadened the gap between the demand for and supply of physician services.

CHANGE UP TO 1983

With the passage of almost a decade and a half, we are in a reasonably safe position to single out the salient developments that occurred to change the urban health scene. For didactic purposes, we will deal first with patients who are able to pay their way or have it paid for them via insurance, and then consider the health care available for the poor and the near-poor. As far as the insured population is concerned, it enjoyed substantial gains:

— There was a striking increase in the number of physicians per 100,000 population, an improvement of almost 50 percent over the ratio existing in the early post-World War II years.

— New hospital beds were added to the system. More personnel were added to hospital staffs, and the average length of stay was reduced, with the consequence that in most cities patients had little or no difficulty in gaining access.

— The passage of Medicare assured most of the elderly, excepting those that had the misfortune to encounter a serious illness, that they would no longer face the threat of impoverishment or have to look to their children for support in the event that they became ill and required one or more spells of hospitalization. Without question, the much improved access of the elderly to the health care system was the most important advance achieved during the last decade and a half.

— While nursing home care began to expand prior to the passage of Medicare and Medicaid, the new inflow of governmental funds after the two amendments of the Social Security Act (XVIII and XIX) further stimulated the growth of nursing homes largely under proprietary auspices. Since Medicare never provided broad insurance coverage for nursing home care, the middle class was not significantly better off than it had been prior to the legislative enactment when it came to such care. Such persons had to pay their own way and would be assisted only after they had consumed all of their resources. At that point, a nursing home patient was eligible to go on to the Medicaid rolls. While some, probably a considerable number, of middle-class patients avoided the letter and the spirit of the law by transferring their assets ahead of time to members of their family, nursing home care did not become broadly available to most patients who sought to pay their own way.

— Since most Medicare beneficiaries also become enrolled in Supplementary Health Insurance (Medicare B, to cover the charges of physicians), the preexisting barriers that had earlier led many of the elderly to ration their visits to physicians were removed, since the elderly no longer had to pay out of pocket for such visits or, at worst, had to pay only a small proportion of the bills, since Medicare covered the bulk.

— The elderly who entered a hospital for treatment had a semi-private or private room, since care in the old-fashioned large wards was not reimbursable under Medicare regulations.

— The combination of Medicare A and B coverage resulted in the shift of many of the elderly from public facilities to voluntary hospitals and to private physicians. They were welcomed by the latter and in turn preferred to be treated in the private sector, believing that they would get better and more personal treatment.

— At the same time that Medicare made the elderly attractive to private health care providers, nonprofit and commercial insurance expanded the sales of major medical insurance policies to persons below and above 65 years of age (Medigap), which resulted by the end of the 1970s in third-party payers — government plus insurance — covering over 90 percent of all hospital costs or charges and about two-thirds of all physician charges. One of the unexpected consequences of this greatly improved coverage was a steep rise in hospital and other medical costs that put great pressure on the whole system, because health care costs kept running consistently several percentage points above the general rise of prices. Hence the mounting concern with "cost containment." However, the middle class, with minor exceptions, had secured access to a good level of health care and was largely freed from the excessive financial burdens of paying for such care.

The situation with respect to the medical care received by the poor and the near-poor (better defined as those who were covered by Medicaid and those who were not covered) changed as follows:

— A central goal of Medicaid was to bring the poor and the near-poor into the mainstream of American medicine and no longer limit their choice of providers to government institutions set up to cater to their needs or to force them to rely on selective voluntary institutions that had a tradition of providing "charity care."

— The best that can be claimed is that the goal was realized only in part. Medicaid patients in many urban communities did gain ready access to voluntary hospitals for inpatient treatments. The situation with respect to ambulatory care was less satisfactory. Most physicians with a middle-class practice did not welcome having their waiting rooms crowded by Medicaid patients and for the most part avoided treating such patients. That many states reimbursed physicians at a low rate militated against their accepting Medicaid patients. In many cities only a minority of physicians catered to the Medicaid trade and sought to make up in volume for the low unit reimbursement rate.

— In New York City and a few other large metropolitan centers, group practices were formed in low-income neighborhoods usually under an entrepreneurial physician, who hired other physicians, often foreign medical graduates, and concentrated on treating Medicaid beneficiaries. The term "Medicaid mills" came to be applied to these undertakings, an unfavorable appellation suggesting that the quality of treatment was distinctly inferior. Many such mills provided a mediocre level of care.

— One of the untoward consequences of Medicaid legislation was the large proportion of the total expenditures that came to be spent on nursing home care — over one-third. Under federal regulations, states had to provide nursing home care as one of the five basic services, and the rapid expansion of such facilities quickly preempted a large part of the total outlays. In a high-income state, a person-year of nursing home care by the end of the 1970s came to around $20,000.

— In New York City several hundred million dollars of Medicaid funds are used annually to help support low-income persons in their own homes by providing them with housekeeping aides in order to reduce the numbers who have to be institutionalized. Such use of Medicaid funds is the exception, and no definitive studies exist as to the cost-benefit of this approach. Clearly, some of the homebound who require only a few hours of paid help a day are better off remaining at home, but many of those who are severely disabled would probably profit if they could be admitted and cared for in a properly run nursing home, of which, regrettably, there are too few.

— The passage of Medicaid unquestionably removed some of the pressure on local and county governments to maintain public hospitals and clinics that previously provided a large part of basic care to the poor and the indigent, but the inability of Medicaid patients to make their way fully into mainstream medicine meant that these public institutions still had an important role to perform. Also, as many voluntary hospitals came under increasing financial pressure in the late 1970s, they sought to cut back their "charity care" by redirecting many Medicaid and even more non-Medicaid-eligible patients to municipally supported institutions.

— If one ascribes, as one must, much of the inflation of medical care costs to the growth of Medicare and Medicaid, then one must recognize that the protective actions taken by many voluntary hospitals to reduce their "charity care" made it increasingly difficult for the near-poor to receive care. They were less likely, now that costs had risen greatly, to be treated by a voluntary hospital.

— Many who are eligible for Medicaid do not remain so over time because of the changes in federal/state criteria or because of changes in their own economic circumstances. For many of the poor, such frequent shifts in eligibility introduce another major impediment to establishing or maintaining a regular source of health care.

— In the period following 1965, the federal government made considerable funding available for the establishment or expansion of community health centers, which had as a primary objective the provision of ambulatory care, often combining preventive and therapeutic services under one roof, located in the middle of a low-income neighborhood. Ancillary aims of these new centers were to provide opportunities for many of the unemployed or underemployed neighborhood population to obtain work in these centers and, wherever possible, to obtain additional training to advance into paraprofessional positions. The federal, state, and local planners greatly overestimated the amount of funding that would become available for the expansion of these centers, but in total the centers provided ambulatory care for several millions of poor persons in the inner city. Regrettably, most of the centers have been unable to develop revenue sources to become financially self-sufficient. As health care funds have become constrained, many of these centers are increasingly at risk.

— The data are clear that the poor and the near-poor have under Medicaid and related governmental efforts substantially increased their access to a real use of health care, both ambulatory and inpatient. Prior to the new legislation, the utilization levels for the urban poor were considerably below those for middle-class persons, but more recently these levels have become equal, or, in many instances, higher. Two caveats: Many of the poor are in worse health than their affluent neighbors and therefore require more health care, and their straightened financial circumstances imply that they often lack the ability to purchase other goods and services that would contribute to improving their health, such as special foods or suitable housing.

SOME CHANGES THAT LOOM AHEAD

As of the beginning of 1983, we know what the president has proposed by way of budget recommendations, but we are still in the dark as to the actions that Congress will take when it finalizes the budget, and much the same uncertainty attaches to the actions of most of the state legislatures, which are also under considerable pressure to slow their outlays and which will look to reductions in their health care expenditures to make economies. Moreover, business coalitions are increasing in number, and they, in turn, are exploring how private dollars for health care that pass through commercial and nonprofit insurance can be used more productively.

It would require a very optimistic orientation on the part of the analyst to believe that the foregoing efforts at cost containment, if they were to be pursued aggressively for a number of years, would not be reflected in a significant reduction of the quantity and quality of health care services available to the American people. Without trying to outguess first how serious the effort of cost containment will be, one can identify from signs on the horizon some likely developments, which will have differential impacts on the middle class with good insurance and the elderly, the poor, and the near-poor, who are among the most vulnerable. Among the likely developments are the following:

— The middle class will continue to have access to a high level of health care services, but it must anticipate substantial out-of-pocket expenditures for covering the costs of such care.

— The increasing number of physicians and the increasing number of new delivery systems from emergency centers to HMOs will provide the middle class with more choices as to providers and settings of care. To illustrate, many physicians are likely to make house calls; others will have office hours in the evening; and ambulatory surgery will increase. But these and still other developments are almost certain to be associated with increasing the amount of money that the middle class has to pay out of pocket and via insurance to maintain its present level of care.

— Persons on Medicare are likely to be among the most vulnerable to such increased outlays because of the pressures on the federal government to slow its total expenditures for health care, of which Medicare accounts for the predominant share. It may take a few years for Congress to act, but it is difficult to see how it can long delay doing so.

— The elderly may have an option of joining one or another form of prepayment scheme (HMO or other) that may provide more ambulatory care relative to inpatient care than the present system permits. However, it is highly unlikely that there will be any significant public-sector investment in additional nursing home beds, much as they may be needed in many areas.

— In an era of continuing intensified efforts at cost containment, the major groups at risk are the poor and the near poor. There is more than a little evidence that many cities are trying to close down, convert, and sell their public hospitals, which have provided a high proportion of both inpatient and ambulatory care for the poor. Voluntary hospitals have been engaged for some time in redirecting patients unable to pay to public institutions. There is little likelihood that in a period of increasing financial pressure voluntary hospitals will add to their charity care. The trend in favor of "preferred providers" (low-bid institutions), as in California, is placing some of the largest teaching hospitals under increasing financial risk.

— In the face of the foregoing, the only kinds of countervailing forces that may operate to protect access of the poor and the near-poor are the success the government may have with the preferred provider approach and the community leadership that some of the business coalitions may demonstrate to ensure that the private sector makes up for some of the cutbacks that are occurring because of retrenchment on the part of government. Unless such a communitywide vantage is adopted, the poor face a reduction, probably a large reduction, in the quantity and quality of health care services to which they will have access.

THE POLITICS OF URBAN
HEALTH CARE

This selective review and forecast of urban health care provides a basis for precipitating a limited number of propositions about the politics that have influenced, if they have not shaped, the health delivery system. Our focus is on the last two decades, a period when the actions of the federal government loomed very large, allowing also for differences in urban environments reflecting intergovernmental arrangements and the division of responsibilities between the private and public sectors. In brief, the following formulations are suggestive of the key political forces that have affected urban health care during the period under consideration and how they are likely to determine the important changes that loom ahead:

— The federal government, through Medicaid, made it very attractive financially (through a 50-83 percent copayment) for states to participate in providing health care coverage for persons on welfare and the medically indigent. After a period of years of accelerating outlays, an increasing number of states have found it necessary to tighten eligibility and to reduce services to relieve the strain on their budgets.

— In a few states such as New York, where cities share with the state the nonfederal costs of Medicaid, the financial arrangements among different levels of government had the result of removing control from the city over a considerable part of its total health outlays.

— The rhetoric about Medicaid bringing the poor into the mainstream of U.S. medical care may have been viewed somewhat skeptically by both politicians and leaders of the poor, but the initial enthusiasm that accompanied the breakthrough of 1965 surely led to rising expectations among the poor for more and better services, to which politicians had to respond.

— The passage of Medicare removed considerable actual and potential financial pressure from city/county governments that, before its enactment, had been faced with increasing demands of the medically indigent elderly for financial assistance or direct health care services, especially hospitalization.

— The relatively liberal reimbursement systems for Medicare, both for hospitalization and for physician services, gave a substantial boost to private-sector providers, except in a few states, where the authorities stepped in to place ceilings on reimbursements to hospitals.

— The federal government, through its sizable expenditures for community health centers and the growing number of categorical health programs, raised the level of interest and participation of neighborhood groups and community-based organizations in the health care arena in which local populations had a twofold concern: with the jobs that became available and with the services that were being offered. As the federal budgets came under increasing pressure, these local groups looked to their cities to make up for the shortfalls of money from Washington, thereby adding to the problems of local officials, whose resources were also increasingly strained.

— The large infusions of federal and state funds for health care services subsequent to 1965 reduced the attention and concern that the city and county officials had earlier directed to maintaining public health institutions, particularly municipal or county hospitals, which had long served as primary providers for the poor and the near-poor. In some cities, such as Philadelphia, a long-established inner-city hospital serving the poor was closed, but the view from 1983 suggests that most large cities will not be able to close down their public hospitals.

— One of the unfortunate by-products of the large-scale increase in public financing of health care was the concomitant weakening of interest on the part of many voluntary hospitals in their community responsibilities. With increasing government funds available to the poor and the elderly to help them pay for essential health care, the private sector's community orientation weakened. It will prove difficult, in the face of the straightened financial circumstances in which most voluntary hospitals find themselves today, to persuade them to respond to the growing plight of the poor whose coverage is being reduced or removed as a result of governmental cutbacks and unemployment.

— Most mayors and county executives have avoided becoming involved over the decades in large-scale efforts to rationalize the relations of public-and private-sector medical institutions, because of the disinclination of the leaders of most voluntary institutions to have their freedom of action and decision making reduced and their potential financial commitments increased. The critical question in the years ahead is whether such arm's-length relationships can be continued in the face of the increasing financial stringency facing both sectors. The answer to this question will greatly affect the directions of health policy in the years and decades ahead.

REFERENCES

Davis, E. and M. Millman (1983) *Health Care for the Urban Poor: Directions for Policy.* Totowa, NJ: Allanheld, Osmun.
Ginzberg, E. and M. Ostow (1983) *Dollars and Service Delivery: The Impact of Federal Funding upon the Health Care System in New York City.* Totowa, NJ: Allanheld, Osmun.
Ginzberg, E. et al. (1971) *Urban Health Services: The Case of New York.* New York: Columbia University Press.
Ginzberg, E. et al. (1983) *Home Health Care: Its Role in the Changing Health Services Market.* Totowa, NJ: Allanheld, Osmun.

10

Urbanization and Health Services:
A World Perspective

MILTON I. ROEMER

□ URBANIZATION IS OCCURRING in all countries. In the industrialized and affluent countries, this has long been obvious, but it is occurring also in the predominantly agricultural and impoverished countries of the world. Cities provide a framework for economic and social development, even when this process is slow and irregular. Insofar as health system organization is one aspect of development, urbanization has created many benefits and also various difficulties for national populations.

This chapter attempts to examine both the positive and the negative impacts of urbanization on health and health services throughout the world. It will approach the issue from the perspective of the overall structure and functions of national health care systems. First, however, a few words are necessary about the nature of cities in the industrialized, compared with developing, countries of the world.

CITIES IN THE CONTEMPORARY WORLD

In the highly developed countries, where industrialization has been occurring for one or two centuries, cities today constitute the centers of civilization. In the early nineteenth century, when the cities of Europe and America were young, urban life meant congestion, slums, communicable diseases, child labor, and all sorts of miseries. But in the twentieth century, systems of sound enviromental sanitation developed, working conditions were greatly improved, social security arose to protect the aged or disabled worker, and school systems moved children from the factory into the classroom.

As cities in the industrialized nations grew older and as modern transportation developed, city dwellers spread outward. The suburbs became the places of residence for the middle class, and the older central-city areas gradually became slums. Even so, with the development of public health services, communicable diseases declined to very low levels. Then, as people lived longer, the chronic degenerative diseases (heart disease, cancer, stroke, diabetes, and so on) became the major causes of disability and death. For many environmental reasons, still not fully understood, the toll of these major killers was not so great among the populations remaining in rural areas. At the same time, thanks to improved sanitation, immunization, and better nutrition, communicable diseases also declined greatly in rural areas. Today, therefore, in America and Europe, the life expectancy (at birth) in the cities is lower and the mortality rates are higher than in the rural areas.

In the developing countries, the formation of cities, and their health implications, have been quite different. Urbanization has indeed occurred, but not to the same degree. Between 1950 and 1970, the less developed regions of the world grew from 15.5 percent to 24.5 percent urban, but the more developed regions grew from 50.8 to 63.9 percent urban. The cities of developing countries were mainly centers of government and commerce, only to a slight degree locales of industry. Comfortable housing was in or near the central city or sometimes in small, select suburban "colonies." Deteriorated housing at the urban center was soon replaced with modern commercial or residential buildings. The most blighted unsanitary areas grew up in the suburbs, in the periurban "belts of misery" — seen today surrounding the large cities of Latin America, Africa, and much of Asia. The squatters in the shacks and shanties of these areas are desperate families who have moved from the rural areas into the urban peripheries in search of employment.

The health implications of these developing-country cities are quite different from those of cities in affluent nations. In spite of the periurban squalor, sanitary conditions are no worse, and are often better than, those in the villages and rural areas. City public health agencies concentrate their efforts on the migrant populations, if only to prevent the spread of communicable diseases to the urban middle classes. Water supply and sewage disposal in the periurban slums may be rudimentary, but they exist. Compared with the primitive environment of the 70-75 percent of developing-country people who are still rural, the conditions of life have improved in the cities. Infant and child mortality from infectious disease and malnutrition still ravage the rural populations. As a result, contrary to conditions in the developed countries, life expectancy in the cities of developing countries is longer than in the rural areas, and mortality rates are lower. Mortality pictures in the large cities of developing countries are, however, beginning to resemble those in the developed countries more each year. Heart disease and cancer are coming to head the list in Kuala Lumpur and Caracas, though not to the same degree as in New York or London. Chronic disease mortality, however, is still not so great as the overall national deaths — not to mention disabilities — from enteric diseases, pneumonia, malnutrition, tuberculosis, malaria, and other scourges of the rural developing countries as a whole.

It is for this reason that the major efforts of the World Health Organization and other international bodies are concentrated on upgrading the living conditions of *rural* people in the developing countries. Urban life, of course, still has plenty of hazards for the poor. Working conditions in small urban enterprises — shops making clothing, furniture, food products, and the like — may be miserable. But urban resources for public health protection and modern medical care are enormously better than those in the rural areas, where the great majority of people in developing countries still live. It is no accident that in the developing countries, the seeds of discontent, rebellion, and revolution, through guerilla movements, find more fertile soil in the rural areas than in the cities.

With this background on the differing health contexts of cities in the developed and developing countries, we may now consider the significance of cities for the operation of health care systems throughout the world.

COMPONENTS OF NATIONAL HEALTH CARE SYSTEMS

The detailed contours of health care systems, of course, differ in all nations. They vary not only with a country's economic development, but also with its social and political ideology. It is possible, nevertheless, to analyze all national health care systems according to a common conceptual framework. Doing so may help us to consider the influence of cities on each component of the health care system's operations.

In very simplified form, a health care system may be analyzed in terms of five major components: (1) development of resources, (2) organization of programs, (3) economic support, (4) management, and (5) delivery of health services. Within each of these components there are several subdivisions or subsystems, the details of which differ enormously among countries, but in one form or another all five components are found in every national health care system. A simple schema of this health system model is shown in Figure 10.1.

The *development of resources* requires social investments for training health personnel, construction of facilities, the manufacture of commodities (including drugs), and the discovery of knowledge. Health personnel, in turn, include physicians, nurses, technicians, and many other disciplines. Even traditional healers in developing countries undergo some preparatory process. Health facilities usually require substantial

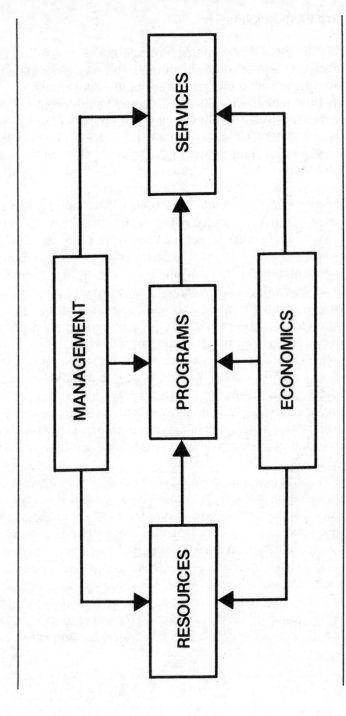

Figure 10.1   Components of a National Health Care System

investment and planning; they include hospitals of many types and also increasing numbers and varieties of health centers, polyclinics, and stations for providing ambulatory care. The manufacture and distribution of drugs and medical supplies are elaborate processes, involving the chemical industry, international trade, advertising, and so on. Knowledge is developed from observation and research, which may be done in all countries, but tends to be concentrated in a few, with dissemination to others through a worldwide literature.

The *organization of programs* is equally or more complex. This may be undertaken by government (at different levels), voluntary bodies, entrepreneurial groups, or other entities. In nearly all countries there is a ministry of health or its equivalent, which is represented in some form at central, provincial, and local levels. Other governmental agencies may be responsible for health insurance programs, occupational health and safety, environmental protection, the health of schoolchildren, and so on. Nongovernmental and nonprofit organizations may tackle certain diseases or promote care for certain populations, like children or the aged. The entire private establishment for providing health services must also be considered part of this system component. While not "organized" in the usual sense, it functions through the mechanisms of a market, in which supply, demand, and price are expected to govern the distribution of services.

Supporting both the development of resources and the organization of programs, there must be a source or *sources of economic support*. For every country there is, in fact, more than one source, and the proportions among the sources have substantial influence on the whole character of a national health care system. Everywhere private households are a source of financing, although these have been increasingly replaced and/or supplemented by insurance (private or statutory), general revenues, charity, even lotteries. General revenues may be derived at different levels and used for diverse purposes. The amounts, as well as the sources, of financing inevitably set limits on the scope and character of a health care system.

A second form of support of the other health care system components is *management* — a short-cut term that encompasses planning, administration, regulation, and evaluation. Planning of the health system may be done with all degrees of completeness, as against dependence on free market operations. Microplanning is always essential for constructing a hospital, but macroplanning may or may not determine the shape of a regional network of hospitals. Administration involves the exercise of authority and assumption of responsibilities in varying balances between centralization and decentralization. Many political and historical influences determine the balance between these poles. Regulation is usually governmental but not always, and it is intended to assure that system performance meets certain standards, especially in its free market sector. Evaluation is always a difficult part of management, dependent usually on a regular flow of information; it enables management to guide health care systems toward achieving their objectives.

Finally, all these components lead to the *delivery of health services* to people. These may be classified in different ways, but a convenient way is in terms of the primary, secondary, and tertiary services. Primary health care includes a wide range of preventive services (environmental and personal), as well as the initial therapeutic response to illness or accident. Secondary and tertiary services usually involve specialization and hospitals. The patterns by which these services are provided differ greatly among countries, involving various degrees of teamwork or organization, various methods of paying personnel, various schemes of recording and communication, and so forth. For certain population groups, such as industrial workers, or certain diseases, such as mental disorder, there may be special patterns of delivery different from the general health care patterns in a nation.

TYPES OF NATIONAL
HEALTH CARE SYSTEMS

The combined characteristics of all five of the foregoing health care system components define the type of system found in each country. In the 160 countries of the world, no two systems are exactly alike, but we can understand better the relationships between urban and rural health services in countries by clustering them into a few main types.

Broadly speaking, there are two principal dimensions along which nations (and their health care systems) may be classified: economic and political. The economic dimension may be scaled with relative ease, by reference to a country's gross national product (GNP) per capita. The industrialized countries tend to be wealthier and the agricultural countries tend to be poorer (with about a dozen oil-exporting nations being exceptional in having high wealth with little industrial development). The political dimension is more difficult to scale, but most observers would agree that there is a range of degrees of sociopolitical organization among nations. At one pole there are countries with a highly permissive, individualistic, and laissez-faire political ideology. At the other pole, there are the highly organized, collectivistic, and socialistic countries. Between the two, there are many countries combining certain features of both; they are moderately organized, they have a relatively highly developed public sector for welfare services, yet their economic systems are essentially capitalistic. The range varies between a maximum of free market dynamics (with only limited planning) and a minimum of market dynamics (with highly centralized planning). For the sake of brevity, we may refer to these three political types of country as entrepreneurial, cooperative, and socialist.

Beyond these two dimensions, economic and political, by which countries and their health care systems may be classified, there are other social forces that obviously play a part. They include historical developments (e.g., previous colonial

ECONOMIC	POLITICAL		
	ENTREPRENEURIAL	COOPERATIVE	SOCIALIST
INDUSTRIALIZED			
AGRICULTURAL			

Figure 10.2 Types of National Health Care Systems

status), religion and culture, climate and geography (tropics, desert, and so on), and more. To be manageable, however, our examination of urban-rural relationships in health care systems will have to be confined to the economic and political dimensions. Describing these types of health care systems, along the two dimensions, gives us a simple matrix, as shown in Figure 10.2.

In all six types of health care systems shown in Figure 10.2, there are contrasts and interrelationships between the cities and the rural areas, with respect to all five of the components of health care systems reviewed earlier. In a general way, the cities have more resources, yet these resources are to some extent a benefit to the rural areas. The cities have more developed organized health programs (serving the poor), yet they also have a larger free market of health service that favors the rich. The cities are the headquarters of health system management, the impacts of which extend far beyond their borders into the rural areas. The cities are sources of the greatest national wealth, some of which supports the costs of health care in the rural regions. Finally, the cities are the place of delivery of all types of health services — primary, secondary, and tertiary — which to some degree serve the whole population of a country. With respect to each of these system components, we may examine more closely some of the urban/rural health problems observable in the current world.

DEVELOPMENT OF HEALTH RESOURCES

In all types of countries, there is a concentration of health personnel, especially physicians, in the large cities, and consequent inadequacies in the rural areas. Medical schools are located in the main cities, where large tertiary-level hospitals serve patients, as well as teaching purposes. In many large cities there are several medical schools; the University of London, for example, has twelve different medical faculties, each attached to a separate hospital. Cities are also the home of most schools for training other health personnel. The main exceptions are the newer training centers for auxiliary "community health workers," which are deliberately being established in rural regions of developing countries, where these personnel will be working.

After their training is completed, a disproportionately large share of doctors and other health personnel settle in the large cities of all countries. It is common for 50 to 60 percent of physicians to become located in the capital city of a developing country, where hardly 10 percent of the population may be living. In industrialized countries the disparities are similar, though less severe. As a result, many national strategies have been applied to attract physicians to rural areas — paying them higher salaries or guaranteeing certain income levels, providing them attractive housing and working quarters, and in many countries mandating a period of rural service for one to three years after medical qualification. More aggressive policies, for achieving greater accessibility of doctors in rural areas, tend to be implemented in the cooperative and socialist countries than in the entrepreneurial countries (as delineated in Figure 10.2).

It is reasonable to expect a higher degree of specialization in medicine in large cities, and this is true everywhere. In most countries (the United States being the major exception), medical and surgical specialists are usually salaried employees of large general hospitals, to which patients with complex disorders may be sent from surrounding regions. Sometimes the urban hospital specialists travel to the smaller peripheral hospitals for consultation and educational purposes. The laboratories of the urban hospitals may perform examinations on specimens sent from small peripheral facilities.

In developing countries, it is common for large investments to be made in one or two magnificent general hospitals in the national capital, while most rural areas with 75 percent of the population lack even simple health centers. In Kenya, for example, some 34 percent of the Ministry of Health budget goes to support one large hospital in Nairobi. Much of this

expenditure goes for the purchase and maintenance of sophisticated medical equipment that may benefit only a few patients.
This type of imbalance has led the World Health Organization
(WHO) to call for "appropriate technology" as more likely to
achieve a minimum level of primary health care for whole
national populations.

At the same time, cities also serve as the heart of a program
for distributing various resources needed throughout a country.
Drugs may be manufactured in cities, or, if they are imported,
they are held in central urban storehouses. In developing countries, where ministries of health operate national networks of
hospitals and health centers, allotments of drugs and medical
supplies are usually sent out periodically. (Unfortunately, the
high costs of imported drugs and transportation problems often
result in shortages at the rural facilities.) Even for private
pharmacies, intermediary agents in the cities customarily dis-
tribute drug supplies throughout a country. To simplify the
problems of drug distribution, WHO has been promoting the
use in developing counries of a relatively short list of two or
three hundred "essential drugs" adequate for meeting some 90
percent of health needs. In both the entrepreneurial and the
cooperative countries, the marketing of thousands of drugs,
many with identical pharmacologic action but different names
and prices, complicates the provision of medical care for both
doctors and patients.

The cities are also the centers of medical research, out of
which new knowledge is derived. Journals or bulletins for the
dissemination of knowledge are sent out from the cities, and
continuing education programs are usually offered at urban
centers. Thus for all types of basic health resources, cities are
the main places of origin, or, with respect to hospital care, they
are the centers for receipt of complex cases. The major issue in
all types of countries, but especially the developing countries,
is to achieve a better balance between the health resources
available in the cities and the rural areas. In the cooperative-
type countries, the operation of a national health insurance
program, and in the socialist-type countries the operation of

public medical service, provide economic leverage for attaining such a balance somewhat more successfully. No country, however, has achieved the level of urban-rural equity that its health leaders advocate.

ORGANIZATION OF HEALTH PROGRAMS

For the most organized programs to provide health care, the cities of the world have natural advantages. With respect to water supplied and sewage disposal, this is manifest. Concentrations of populations make it economically and structurally feasible to deliver clean piped water to households and likewise to develop sanitary sewerage networks. At the same time, the hazards of malfunctioning sanitary systems in cities are great, insofar as contamination of water lines at one point may spread disease to thousands of people. Constant vigilance, of course, is necessary.

The disposal of urban sewage has become an increasing problem as cities have multiplied and grown larger. When towns were small, sewage could be discharged innocently into a nearby river and soon be rendered innocuous by biological processes. But with huge quantities of waste to dispose of, large cities must now have elaborate sewage treatment plants for removing infectious solid materials before the liquid effluent can be discharged into a body of water. Similar and often more serious problems concern the disposal of toxic industrial waste. In the highly industrialized countries, regulatory legislation to prevent stream pollution has become increasingly necessary, even though it adds a great deal to the cost of maintaining city life and of carrying out industrial production. The disposal of solid waste or garbage also presents monitoring problems in the cities, requiring special schemes (e.g., sanitary landfills) to circumvent the creation of public nuisances.

The problems of rural environmental sanitation are very different in the industrialized countries from those in the agricultural countries. In the small towns or even villages of the more developed countries, public systems for both water and sewage are usually quite feasible; isolated farmhouses can have safe wells and septic tanks. In the less developed countries, potable and piped water is a rarity outside of the main cities. In the villages, there may sometimes be a well producing safe water in the village square, where women come to collect it in large jugs. Excreta disposal is through latrines for one or more dwellings, or there may be no special equipment at all. The vast toll of infant mortality in the developing countries is caused largely by enteric diseases resulting from deficiencies in environmental sanitation. Even in the large cities of the developing countries, where public water and sewerage systems operate, the maintenance of these systems is usually beset with problems

Programs for personal health service have corresponding advantages in cities compared with rural areas. Maternal and child health clinics are ordinarily much more accessible to mothers in a city than in a rural area. Public transportation is usually available in cities, in addition to which traveling distances are shorter. Rural families tend to have more children, but services are harder to reach. Many studies in developing countries have shown the utilization of health center services to decline as the distance from households to the center increases. The same applies, of course, to clinics for tuberculosis, general primary health care, or other purposes.

With their large congested populations, vehicle traffic, and factories, serious accidents are a major problem in cities everywhere. In the highly developed countries, emergency medical services (EMS) programs have been developed in many cities. Ambulances, special communication arrangements, and hospital emergency departments are organized for rapid response to any call for help. These EMS programs respond to cardiac crises as well as traumas. In the socialist countries, these programs have been developed to a very high level. Some developing countries have organized EMS programs in their capital cities, but seldom elsewhere. The people of rural areas in all types of countries may suffer serious accidents in agricultural work, mining, or on the roads, but emergency responses are seldom available.

Health programs requiring voluntary initiative of groups of people are also mainly products of the city. Societies to help victims of tuberculosis, to fight cancer, to rehabilitate crippled children, to provide home nursing services to the chronically ill, and so forth are formed where large numbers of people with similar problems can interact. In many developing countries and also in the cooperative-type countries, voluntary health agencies of this sort may even be subsidized by government. They seldom operate, however, in the rural areas.

Specifically urban, of course, are programs of occupational safety and health. In the industrialized countries, particularly of the cooperative type, these protective services have been developed to a high level. In the socialist industrialized countries, these programs are also well developed, and they usually provide at the workplace general medical care as well as prevention. In developing countries of all types, occupational safety and health programs are quite weak, outside of very large enterprises. Some of the largest of these enterprises — mines, food-processing plants, and the like — may be in isolated rural regions, where the industrial program provides the only health service available.

ECONOMIC SUPPORT ISSUES

In all the industrialized countries of the world, the productive processes occurring in cities constitute the nation's major source of wealth. Even in the developing countries, dependent for their income mainly on the sale of agricultural products, the wealth of families is invariably greater in the cities than in the rural areas. As a source of governmental revenue, therefore, cities usually contribute much more per capita than the countryside. Insofar as public-sector programs for health care are brought to rural people, this means that city wealth brings benefits to rural as well as urban populations.

The degree of this redistributional effect naturally varies with the type of taxation policy in countries. If the scale is highly progressive, as in the cooperative-type industrialized countries, the special benefits of urban taxation for rural areas are great. In developing countries, especially of the entrepreneurial type, the rates of taxation are relatively modest, so that any subsidies for rural areas tend to be meager. Insofar as a developing country's public revenues come from taxes on agricultural exports, as against personal incomes, this whole redistributional effect would be weaker.

One form of economic support for health services brings special benefits to urban people, particularly in developing countries. This is social insurance or social security for financing health care. In the industrialized countries, national health insurance, as it is usually called, has come to protect urban and rural people alike, or nearly so. In the developing countries, however, where statutory health insurance has been enacted — in Latin America, the Middle East, India, and elsewhere — it is nearly always restricted to persons getting regular wages or salaries. This means essentially industrial, commercial, or government workers, who are largely city dwellers. Only to a small extent — for example, in Mexico or Costa Rica — have these programs been extended to agricultural workers or families.

Statutory health insurance in developing countries differs from that in developed countries in another important respect. The health insurance agency typically engages its own doctors and other personnel on salary and operates its own hospitals and clinics, unlike the traditional European pattern of using insurance to pay for services rendered by private practitioners and hospitals under various auspices. This means that city people covered by these programs in developing countries have not only the financial protection but also the assured availability of the health personnel and facilities to provide the needed services. Some have argued that these health insurance programs in developing countries aggravate the disparities between urban and rural health services. One finds, however, no greater disparity in countries with such insurance than in countries without it. In reality, statutory health insurance probably mobilizes more efficiently the health expenditures of urban people, which would otherwise have been made in the private medical market.

Purely private household spending for health care is another source of economic support that is appreciably stronger in cities than in rural areas, in both industrialized and agricultural countries. Insofar as family incomes in the cities are nearly always higher than in the rural areas, these personal expenditures naturally attract greater medical care resources and services. With physicians, dentists, hospital beds, pharmacies, and so on, much more accessible to urban populations, rural populations in developing countries must depend for their medical care far more on untrained traditional healers. The entire role of private spending, of course, is much greater in entrepreneurial-type countries, of any stage of economic development, than in cooperative and socialist countries. In fact, the hallmark of the health care systems of the latter types of countries is their substantial replacement of private spending with social forms of health care financing.

Still another form of economic support for health services is charity or voluntary donations. This, too, tends to be more abundant in cities, where higher family incomes make possible such generosity. Sometimes the church or church-related organizations mobilize charitable contributions for building hospitals, financing certain services for the poor, supporting medical research, and the like. Voluntary health agencies focused on certain diseases, such as tuberculosis or cancer, are typically able to raise much more money in urban than in rural populations. In many countries, charitable societies raise funds for hospitals by the sale of lottery tickets, once again primarily in the larger cities. In socialist countries, this form of health care financing has declined to negligible proportions.

MANAGEMENT OF HEALTH CARE SYSTEMS

Every national health care system and subsystem requires management — that is, the exercise of some sort of leadership or guidance to keep it moving toward its objectives. Broadly speaking, management in health care systems includes planning, administration, regulation, and evaluation. The effects of these processes are felt everywhere in a country, but the central source or sources from which they emanate are ordinarily the capitals of nations and/or their constituent provinces.

The planning of health resources and services has come to be accepted as necessary in almost all modern nations. While once identified only with Soviet Russia, since World War II all countries — and particularly the developing countries — have come to recognize the inequities and inefficiencies resulting from purely free market dynamics. As political movements have come to establish health care as an entitlement or social right, governments have taken more and more deliberate actions to plan the distribution of health personnel, facilities, equipment, drugs, and programs for prevention or treatment in accordance with needs, rather than simply market demands. In most countries there are ministries or agencies for overall planning purposes, and health is typically included; ministries of health also usually include planning offices. Such activities are invariably carried out in cities, but a major thrust tends to be toward helping rural areas overcome their handicaps. The strength of health planning is, of course, greater in the socialist and cooperative than in the entrepreneurial countries.

Administration, as a part of management, refers to the whole process of policy formulation, decision making, delegation of responsibility, communication and information, coordination, and so on. In every health care system there is a certain balance between centralized and decentralized responsibility with respect to all of these matters. The less highly developed countries tend to be more centralized, displaying little trust in the capabilities of local communities. In the more highly developed countries, administration is usually much more decentralized and localized. Once again, the main cities are the seats of power in the developing countries (with three-fourths of the world's population), and a greater sharing of responsibility between the urban center and the rural periphery occurs as countries develop. Centralized authority also tends to be greater in the socialist and some of the cooperative-type countries, although variations may still be wide — for example, between the health care systems of the USSR and China.

Regulation may be governmental or voluntary, but its main purpose is to assure that health resources and services meet specified standards of quality. Because of the complexity of medical and other health services, one can hardly expect the average consumer to assess a health care product effectively. Rules and regulations are established in all countries to protect the consumer against abuses or simply deficiencies of health services in the free market. Such regulation is ordinarily based on standards that are developed in the cities, but they must be enforced everywhere in a nation. All too often, regulatory standards may be very well formulated, but the staff and resources to enforce them throughout a country are inadequate. The weakest enforcement of health standards tends to be in isolated rural areas. Regulation tends to be greater, of course, in cooperative and socialist countries, where health service has been more extensively assured as a social responsibility.

DELIVERY OF HEALTH SERVICES

The final component of national health care systems is the pattern by which the services are provided to people. By reason of the physical and social nature of city life, the urban pattern is usually quite different from that in rural areas.

With regard to ambulatory health care, in most countries there are many private medical practitioners among whom the city dweller may choose. In addition, for lower-income people there are often health centers or clinics, where primary health care or categorical care (for expectant mothers, small children, patients with venereal disease, and so on) may be obtained. In rural areas everywhere, the choices are fewer. Private medical or dental offices are relatively few, and organized health centers, with teams of health personnel, are not so readily accessible. It is only in the fully socialist countries, such as the USSR, Cuba, or China, that patterns of ambulatory care delivery in both cities and rural areas are essentially limited to health teams working in organized polyclinics or health centers. However, in all countries, both industrialized and developing, ambulatory health care is becoming more frequently delivered in the organized setting of community health centers, group practice clinics, or the like, in contrast to solo medical practices.

Patterns of hospital care delivery in most of the world involve "closed staffs" of salaried doctors and other personnel, both in cities and in rural areas. This applies to both governmental and voluntary, nonprofit hospitals, although not to proprietary hospitals. The "open staff" hospital, where almost every local physician may treat patients, found in the United States and Canada, is quite exceptional in the world scene. Large multispecialty hospitals are, of course, located in the cities everywhere, while rural areas are served by smaller general hospitals staffed and equipped to handle relatively uncomplicated cases.

Developing regionalized networks of hospitals is a policy that has been spreading around the world, with the objective of making appropriate hospital care available to all people, wherever they may live. In effect, a regionalized hospital system makes tertiary-level hospital care as available to the most isolated country dweller as to the residents of a large city. Regionalization requires an efficient referral process from the rural periphery to the urban center and a flow of consultation services from the urban center to the rural periphery. It is, in effect, a strategy for bringing the benefits of urban medical resources to the rural population, at least for the care of conditions serious enough to require hospitalization. In

entrepreneurial countries, where each hospital is largely auton-
omous, the regionalization concept has been less fully implemented than in the cooperative or socialist countries, where
hospitals are regarded as part of a national system.

With respect to health care delivery, there are special patterns for the provision of certain services, such as psychiatric
care, rehabilitation of the seriously disabled, or diagnosis and
treatment of occupational diseases. Once again, larger cities
are likely to have resources for such services that are not to be
found in small towns or rural areas. So long as the capacities of
these programs are great enough to handle referrals from the
surrounding region, the city can be a source of nationwide
health benefits rather than special privileges.

FINAL COMMENT

This review of urban health services around the world may
be enough to demonstrate that, in spite of the problems of
urbanization — difficulties in housing, education, transportation, environmental pollution, and so on — the social developments possible in cities have probably created greater
benefits than difficulties in most nations. It is the rural areas
that more often suffer the handicaps. These apply to all the
major components of health care systems — the production of
resources, the organization of programs, economic support,
management, and the delivery of health services. With respect
to all these processes, the cities have been capable of greater
progress and the rural areas have depended largely on the cities
for their advancement.

The equilibrium between cities and rural areas obviously
differs among countries. The differences apply along both
economic and political dimensions. In countries with greater
economic development, the relative advantages of urban over
rural life are not so great; indeed, the difficulties of urban life
may in some ways outweigh their benefits, relative to life in the

countryside. In the economically less developed countries, the larger cities have nearly all the advantages. Underdevelopment in education, housing, nutrition, employment, culture, health — almost everything — is mainly rural. The social development of rural populations is largely dependent on movements originating in the cities.

Along the political dimension, the urban-rural relationships are more complex. With respect to social organization of all the processes constituting a national health care system, a greater degree of such organization tends to achieve greater national solidarity and equity. National planning, national financing, national standards, national programs of health service tend to disseminate the benefits of urbanization for the welfare of everyone. There are obvious dangers, however, in overcentralized authority, which only the alert participation of local community people in policymaking can prevent. The challenge for achievement of health care equity — between urban and rural people, between rich and poor, between young and old — is to strike the optimal balance between social responsibility, on the one hand, and individual freedom and motivation, on the other.

REFERENCES

Basch, P. E. (1978) *International Health*. New York: Oxford University Press.
Douglas-Wilson, I. and G. McLachlan [eds.] (1973) *Health Service Prospects: An International Survey*. London: Lancet and Nuffield Provincial Hospitals Trust.
Elling, R. H. (1980) *Cross-National Study of Health Systems: Political Economies and Health Care*. New Brunswick, NJ: Transaction.
Roemer, M. I. (1976) *Health Care Systems in World Perspective*. Ann Arbor: Health Administration Press.
Roemer, M. I. (1977) *Comparative National Policies on Health Care*. New York: Marcel Dekker.
Sidel, V. W. and R. Sidel (1977) *A Healthy State: An International Perspective on the Crisis in United States Medical Care*. New York: Pantheon.

About the Contributors

TOM ANDERSON is a research assistant at the Urban Research Center and a graduate student in urban affairs at the University of Wisconsin — Milwaukee. He specializes in health services research and is particularly interested in problems of coordination of services for the chronically ill. His current research effort concerns the economic implications of the spatial differentiation of health services in urban environments.

H. DAVID BANTA, a former assistant director of the Health and Life Sciences Division, Office of Technology Assessment, U.S. Congress, is the deputy director of the Pan American Health Association. He holds an M. D. from Duke University and master's degree in public health and in health services administration from Harvard University's School of Public Health. He is particularly interested in the benefits, risks, and costs of medical technology and generally in the use of scientific knowledge in public policymaking.

CLYDE J. BEHNEY is a doctoral candidate in the School of Government at George Washington University. He joined the staff of the Office of Technology Assessment, U.S. Congress, in 1977. He served as staff to OTA projects on saccharin and on the efficacy and safety of medical technology, directing the latter study after the original director left. He has directed projects on cost-effectiveness of medical technology, on technology transfer at the National Institutes of Health, and on technology and handicapped people. In 1981 he was appointed program manager for health.

ANNE KESSELMAN BURNS is currently project director of the study Medical Technology and Cost of the Medicare Program, at the Office of Technology Assessment, U.S. Congress. She was a key member of the project staff for the recently completed OTA project on technology and handicapped people. Prior to joining the OTA staff, she was assistant director of the Community Hospital Program of the Robert Wood Johnson Foundation. A graduate of Princeton University, Burns earned a master's degree in health administration from the University of Massachusetts School of Public Health.

ELI GINZBERG, Hepburn professor emeritus of economics and special lecturer in business, health, and society, is director of the Conservation of Human Resources and Revson Fellows Program on the Future of the City of New York, Columbia University. He is a member of the Institute of Medicine and the American Academy of Arts and Sciences; winner of the Blue Cross/Blue Shield Fiftieth Anniversary Achievement Award in Medical Economics; and author of over sixty books on human resources and health policy.

FRED H. GOLDNER is professor of sociology at Queens College, City University of New York. His published work includes studies of organizational analysis with a major concern for professionals in organizations, ideology and social organization, and private government. From 1977 to 1978 he was chief-of-staff and executive deputy to the president, New York City Health and Hospitals Corporation.

KAREN GRANT is associated with the Department of Sociology at Boston University.

ANN LENNARSON GREER is an associate professor of sociology and urban affairs and a center scientist of the Urban Research Center at the University of Wisconsin — Milwaukee. She combines an interest in the health services organization and policy with a background in the study of urbanization and urban politics. Her recent empirical research has focused on the diffusion of new medical technologies, governance and administration of health and mental health organizations, and coordination of community services for the mentally ill and the elderly. Her books include *The Mayor's Mandate* and *Neighborhood and Ghetto*. Recent articles appear in *The Social Impact of Medical Technology* (J. Roth and S. Ruzek, eds.), *Policy Sciences, Journal of Medical Systems, The Practice of Policy Evaluation Research* (D. Nachmias, ed.), and *Fiscal Retrenchment and Urban Policy* (J. P. Blair and D. Nachmias, eds.). In recent years she has served as a consultant to the Office of the Assistant Secretary for Health, the Institute of Medicine, and the Health Resources Administration.

SCOTT GREER, distinguished professor in sociology at the University of Wisconsin — Milwaukee, is affiliated with the Urban Research Center and the Departments of Sociology and Urban Affairs. He is author of *The Emerging City, Urban Renewal and American Cities*, and *The Logic of Social Inquiry*. He is a frequent contributor to Urban Affairs Annual Reviews and most recently edited (with R. D. Hedlund and J. L. Gibson) *Accountability in Urban Society*. His general interests are in social organization, social change, and social and cultural integration. His current research is on the roles of popular movements and lay governing boards in the instigation and direction of social change, with particular reference to health services and care of the mentally ill. Publications on such matters appear in *The Practice of Policy Evaluation Research* (D. Nachmias, ed.), *Sociological Quarterly,* and *Policy Sciences*.

SUSAN JENNINGS is senior research associate at Cambridge Research Center, American Institutes for Research. She also is associated with the Department of Sociology at Boston University.

DANIEL M. JOHNSON is professor and former chairman of the Department of Sociology and Anthropology at Virginia Commonwealth University. Prior to his service as department chair, he was co-director of the Ph. D. program on social policy and social work. He is co-author of *Black Migration in America: A Social Demographic History*.

JOHN B. McKINLAY is professor of sociology at Boston University and director of research development at Cambridge Research Center, American Institutes for Research. He also holds appointments at the Massachusetts General Hospital and Beth Israel Hospital. He has published widely in the areas of the distribution of health care and social class, economic aspects of health care, politics and law in health care policy, and clients and organizations. With S. M. McKinlay, he has published "Medical Measures and the Decline of Mortality" and is preparing a volume refuting the notion that the health of the United States is improving.

SONJA M. McKINLAY is associate professor of community health at Brown University and principal research scientist at Cambridge Research Center, American Institutes for Research. Her research and publications include studies of medical measures and the decline of mortality and, with John B. McKinlay, she is preparing a refutation of the notion that the nation's health is improving.

J. JOHN PALEN is an urban sociologist at Virginia Commonwealth University, interested in the consequences of urbanization for other countries. His recent books include *The Urban World; City Scenes; Gentrification, Displacement, Neighborhood Revitalization* (co-editor); and *The Urban Explosion*. He is currently a visiting professor at the National University of Singapore.

MILTON I. ROEMER has been professor in the School of Public Health at the University of California, Los Angeles, since 1962. He taught previously at Yale and Cornell universities. He earned the M. D. degree in 1940 and holds master's degrees in sociology and in public health. He has served at all levels of health administration — as a county health officer, a state health official, an officer of the U.S. Public Health Service, and a section chief of the World Health Organization. He is a member of the Institute of Medicine, National Academy of Sciences. As an international consultant, Dr. Roemer has studied health care organization in fifty-two countries. In 1977 he received the American Public Health Association's International Award for Excellence in Promoting and Protecting the Health of People.

ALAN SAGER has been investigating urban hospital behavior for several years and is now writing a book, *The Closing of Hospitals that Serve the Poor.* He is on the faculty of Boston University School of Public Health and holds a Ph. D. from the Department of Urban Studies and Planning at MIT. Dr. Sager is vice-president of the Health Planning Council for Greater Boston and clerk of the Massachusetts Easter Seal Society. His other research and policy interests are in long-term care. He is the author of *Planning Home Care with the Elderly* (Ballinger, 1983).

HENRY J. SCHMANDT is professor of urban affairs at St. Louis University. He is a former Chairman of the Southeastern Wisconsin Regional Planning Commission and has written extensively in the field of urban government and politics, with particular emphasis on policy issues of concern to the cities.

GEORGE D. WENDEL is professor of political science and, since 1968, director of the Center for Urban Programs, at St. Louis University. For several years, he has participated in evaluative research on federal aid to cities in collaboration with Richard P. Nathan at the Brookings Institution and at the Woodrow Wilson School of Princeton University. Since 1978 he has evaluated neighborhood-based municipal health care delivery with Eli Ginzberg and associates at the Conservation of Human Resources, Columbia University.